FOCUS ON
FRESHWATER
AQUARIUM FISH

FOCUS ON
FRESHWATER
AQUARIUM FISH

Geoff Rogers • Nick Fletcher

FIREFLY BOOKS

A FIREFLY BOOK

Published by Firefly Books Ltd. 2004

Originally published in 2004 by
Interpet Publishing

Copyright © 2004 Interpet Publishing

First printing

Publisher Cataloging-in-Publication Data (U.S.)

Roger, Geoff.
Focus on aquarium freshwater fish / Geoff Rogers ;
Nick Fletcher. —1st ed.
[208] p. : col. photos. ; cm.
Includes index.
Summary: A photographic book of aquarium
freshwater fish listed by common and Latin name
and divided by family groupings with details on the
size and characteristics of each.
ISBN 1-55297-936-9
1. Aquarium fishes — Pictorial works.
I. Fletcher, Nick. II. Title.
639.31   dc21   SF457.R64   2004

**National Library of Canada Cataloguing
in Publication**

Rogers, Geoff, 1944-
Focus on aquarium freshwater fish
/ Geoff Rogers, Nick Fletcher.
Includes index.
ISBN 1-55297-936-9
1. Aquarium fishes — Identification.
2. Aquarium fishes — Pictorial works.
I. Fletcher, Nick, 1949- II. Title.
SF457.R63   2004   639.34   C2004-900995-8

Published in the United States in 2004 by
Firefly Books (U.S.) Inc.
P.O. Box 1338, Ellicott Station
Buffalo, New York 14205

Published in Canada in 2004 by
Firefly Books Ltd.
66 Leek Crescent
Richmond Hill, Ontario L4B 1H1

Created by Ideas into Print, England
Design/prepress by Stuart Watkinson, England
Production management by Consortium, England
Print production by Sino Publishing, Hong Kong
Printed in China

Taking these photographs has been a challenge and a pleasure. The challenge has been mostly physical — sitting crouched in front of the tank for hours — and sometimes mental — from the anguish of waiting for some fish to do anything interesting! The pleasure kicks in when the photographs succeed in our aim of showing aquarium fish as living creatures completely at home in an environment quite different from our own. A thin sheet of glass separates our two worlds and yet it is possible to capture the essence of their lives as if no barrier exists between us. This book gives us the opportunity to look more closely at a wide range of fish that many of us glimpse only fleetingly in dealers' tanks. Of course, the fish know that I am there with my digital camera pointed in their direction. The cichlids are particularly curious — look at their eyes following my every movement! And some fish, such as the fancy goldfish, seem to relish the opportunity to show off their finery, moving gracefully this way and that so that I can capture their flowing fins from all angles. Of course, I could not have taken these pictures without the fish and I would like to say a huge thank you to all the aquatic stores and fishkeepers who kindly and patiently allowed me to "shoot" their fish. The results are on the page — living snapshots of aquarium fish that are true stars in their own fascinating and vibrant underwater world. **Geoff Rogers**

INTRODUCTION

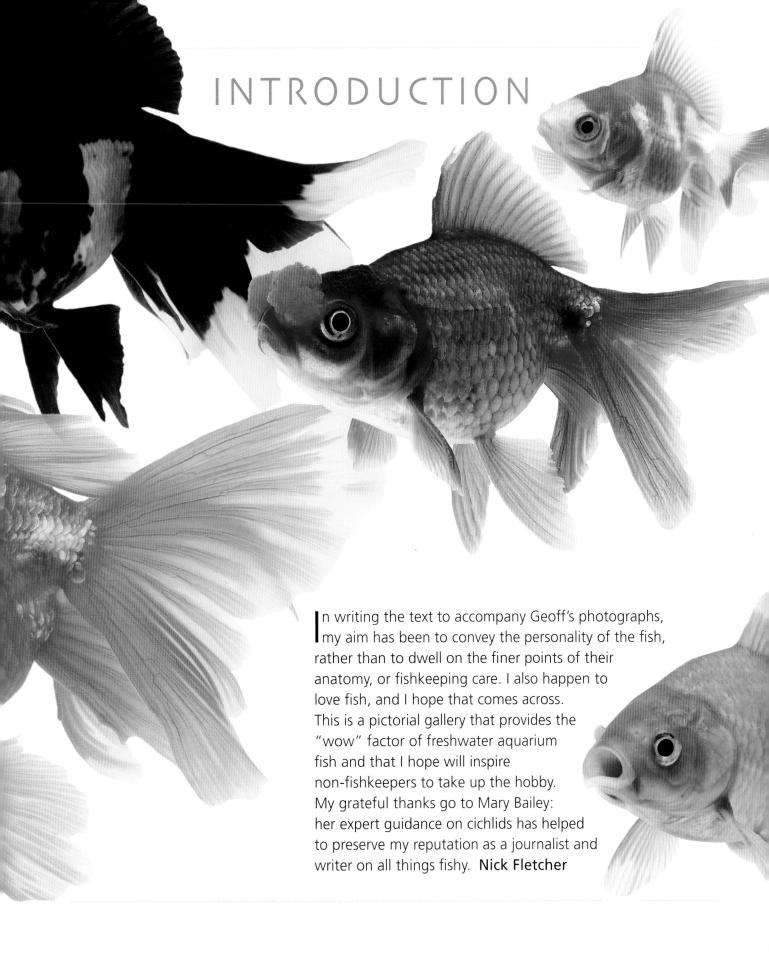

In writing the text to accompany Geoff's photographs,
my aim has been to convey the personality of the fish,
rather than to dwell on the finer points of their
anatomy, or fishkeeping care. I also happen to
love fish, and I hope that comes across.
This is a pictorial gallery that provides the
"wow" factor of freshwater aquarium
fish and that I hope will inspire
non-fishkeepers to take up the hobby.
My grateful thanks go to Mary Bailey:
her expert guidance on cichlids has helped
to preserve my reputation as a journalist and
writer on all things fishy. **Nick Fletcher**

CONTENTS

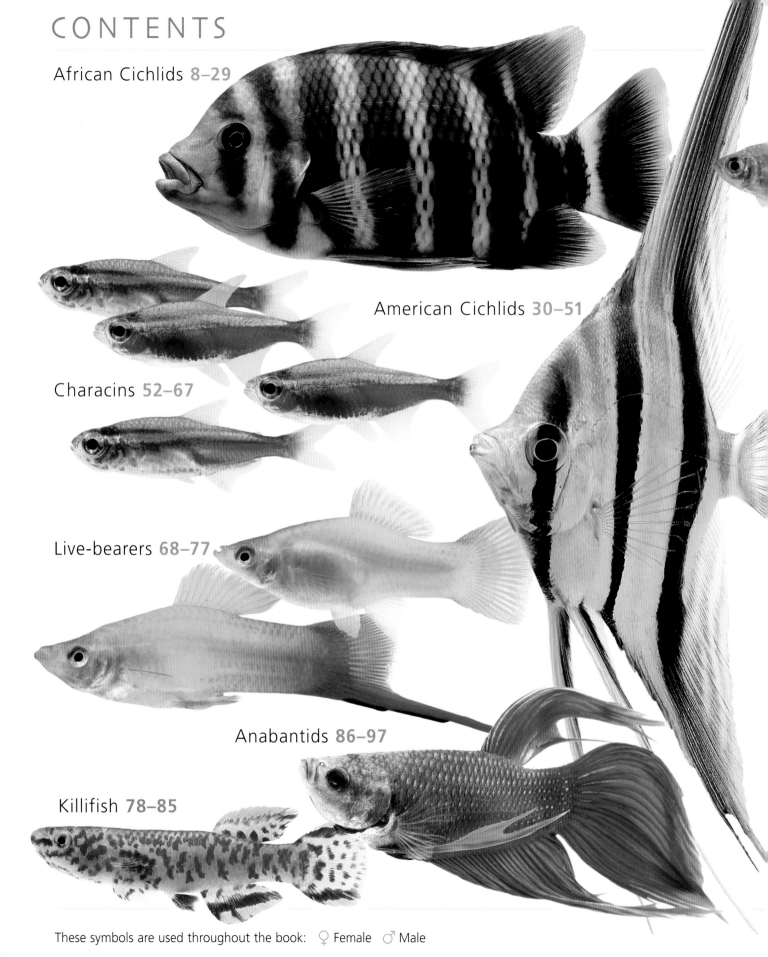

African Cichlids 8–29

American Cichlids 30–51

Characins 52–67

Live-bearers 68–77

Anabantids 86–97

Killifish 78–85

These symbols are used throughout the book: ♀ Female ♂ Male

Danios, Minnows & Rasboras 98–103

Catfish 104–131

Rainbowfish 132–137

Loaches & Sharks 138–149

Barbs 150–157

Goldfish 174–203

Oddballs 158–173

Index & Acknowledgments 204–208

Africa enriches the hobby with a vast array of cichlids that, like their homeland, are beautiful, mysterious and challenging. There is something here for everyone.

Cichlids, more than any other fish, have the collective ability to occupy all available habitats — from forest streams to the giant East African inland seas known as the Great Rift Lakes (Lakes Malawi and Tanganyika). Here, along rocky or sandy shorelines and in every level of open water, are cichlids perfectly adapted to feed and breed in very specific biotopes.

The Malawi cichlids of most interest to hobbyists are the Mbuna, which graze *Aufwuchs* (algae) and associated creatures from the surface of rocks. Mbuna do not stray far into open water, and so single species have evolved into distinct geographical variants tied to their location around the lake. The genera *Pseudotropheus, Labeotropheus,*

Labidochromis, Melanochromis and *Metriaclima* are commonly imported for the hobby. There are also sand-dwellers, such as *Aulonocara, Lethrinops* and many haplochromines; and the Utaka (*Copadichromis* and *Placidochromis*), which are typically found in more open water, and feed on plankton.

Almost all Malawi cichlids are maternal mouthbreeders. Permanent pairs are not formed — each male spawns with all available females. The "egg-dummy" markings on the males' anal fins were once thought to encourage females to pick up sperm by mouthing in the right area, but are now

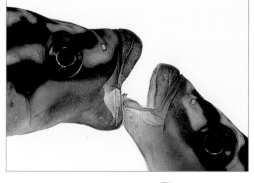

Julidochromis sp. 'Gombi' mouth sparring during pair-bond formation.

considered merely an attractant, males with the most spots being preferred.

Lake Tanganyika cichlids are arguably even more diverse: from the pair-forming, cave-spawning *Julidochromis*, to the mouthbreeding shoals of delicate *Cyprichromis* and their main predator, *Cyphotilapia frontosa*, which hunts at dawn and dusk, when the prey is

Tilapia buttikoferi — not a great social mixer.

CICHLIDS

A popular Mbuna — Metriaclima *(formerly Pseudotropheus)* zebra.

Jewel Cichlid, a dweller of rivers and streams.

least active. There are tiny shell-spawning *Lamprologus* species, Goby cichlids, such as *Eretmodus cyanosticus*, that live in rock rubble along the shore, and sand-dwelling *Xenotilapia*, in some species of

which both sexes mouthbreed the eggs. Topping the food chain is *Boulengerochromis microlepis,* at 32 inches (80 cm), the largest Rift Lake cichlid of all.

West Africa's most popular aquarium cichlids belong to the genus *Pelvicachromis*, best-known in the hobby being *P. pulcher*, the "Krib." Small and relatively peaceful, they can be kept in a community aquarium and will spawn in a flowerpot cave or inverted coconut shell. Do not

neglect rapids cichlids from the Zaire (Congo) River. Members of the genus *Steatocranus*, including *S. casuarius* (African Blockhead), have reduced swim bladders and live on the bottom in fast currents. They, and the similarly adapted *Teleogramma brichardi*, are cave-spawners.

There are larger riverine cichlids to consider, too: the Jewel Cichlids *(Hemichromis* spp.*)* and the boldly striped *Tilapia buttikoferi*.

COMPRESSICEPS

Altolamprologus compressiceps

Mature size: Males 6 inches (15 cm)
Females 3.5 inches (9 cm)

Characteristics: A doleful-looking character from Lake Tanganyika, not cut out for a life of domestic drudgery. The male (shown here) defends nest sites and mates with several females in turn, yet plays no part in brood care. In fact, he would eat the fry if he could get to them. Young fish make up his diet, and he has been known to egg-rob the nests of other cichlids. But nature has a clever strategy: Compressiceps exploit rock crevices and fissures as ideal spawning sites. The male is much larger than the female, and so cannot gain access to the eggs—all he can do is fan his sperm into the hideaway. His wives, on the other hand, can squeeze right inside, plugging the entrance with their bodies. Both sexes are laterally compressed, a characteristic reflected in the scientific name. This means they can pursue prey in among the rocks.

BAENSCH'S PEACOCK

Aulonocara baenschi

Mature size: 4 inches (10 cm)

Characteristics: This Lake Malawi sand-dweller is found at one location, so there is only a single strain — although *Aulonocara* species in general are prone to hybridizing in the aquarium. These are the mildest-mannered of Malawi cichlids, so never mix them with Mbuna. The egg-dummy spots are probably just decorative.

BLUE REGAL
PEACOCK

Aulonocara sp. 'Nagara'

Mature size: 4 inches (10 cm)

Characteristics: This fish is probably a regional strain of *A. stuartgranti*, named after a famous fish-collector and exporter on the shores of Lake Malawi. The lake is so large (365 miles/ 584 km long and 52 miles/83 km wide) that it is effectively a freshwater inland sea, and populations of the same species can show a remarkable diversity, each evolving in a slightly different direction. Females carry up to 50 eggs in their mouths for about 21 days before the fry are released. This fish is a male.

LEPTOSOMA JUMBO

Cyprichromis sp. 'Leptosoma Jumbo'

Mature size: 4.3 inches (11 cm)

Characteristics: These shoaling Lake Tanganyika cichlids inhabit the open waters of the eastern shore around Kigoma. They are plankton feeders and, in their turn, important food fish for some of the lake's larger predators, notably *Cyphotilapia frontosa*, which ambushes them at dawn and dusk, when they are least active.

FRONTOSA

Cyphotilapia frontosa

Mature size: 14 inches (35 cm)

Characteristics: The lazy Frontosa picks off drowsy prey fish at dawn and dusk. These are juveniles; mature specimens develop a head hump.

RED JEWEL CICHLID

Hemichromis lifalili

Mature size: 4 inches (10 cm)

Characteristics: The Jewel Cichlid from the early days of the hobby was the very aggressive West African forest fish *Hemichromis guttatus*. The one now on general sale is *H. lifalili*, not quite such a handful but still extremely territorial at breeding time. Its striking red coloration occurs naturally, deepening still further during courtship.

♀

♂

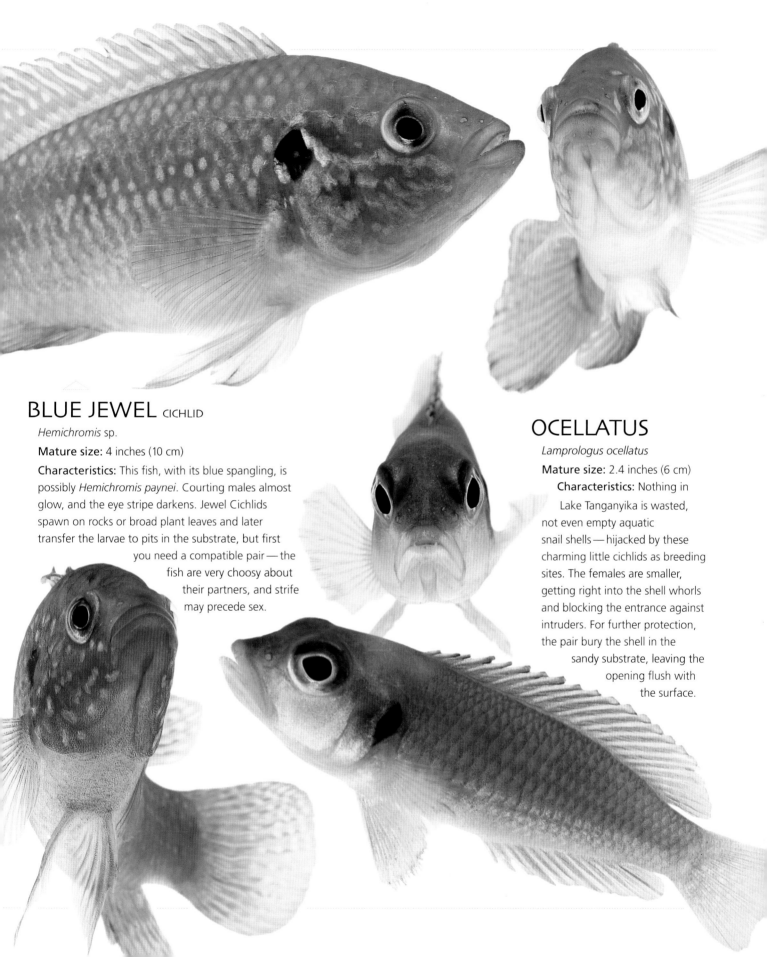

BLUE JEWEL CICHLID

Hemichromis sp.

Mature size: 4 inches (10 cm)

Characteristics: This fish, with its blue spangling, is possibly *Hemichromis paynei*. Courting males almost glow, and the eye stripe darkens. Jewel Cichlids spawn on rocks or broad plant leaves and later transfer the larvae to pits in the substrate, but first you need a compatible pair — the fish are very choosy about their partners, and strife may precede sex.

OCELLATUS

Lamprologus ocellatus

Mature size: 2.4 inches (6 cm)

Characteristics: Nothing in Lake Tanganyika is wasted, not even empty aquatic snail shells — hijacked by these charming little cichlids as breeding sites. The females are smaller, getting right into the shell whorls and blocking the entrance against intruders. For further protection, the pair bury the shell in the sandy substrate, leaving the opening flush with the surface.

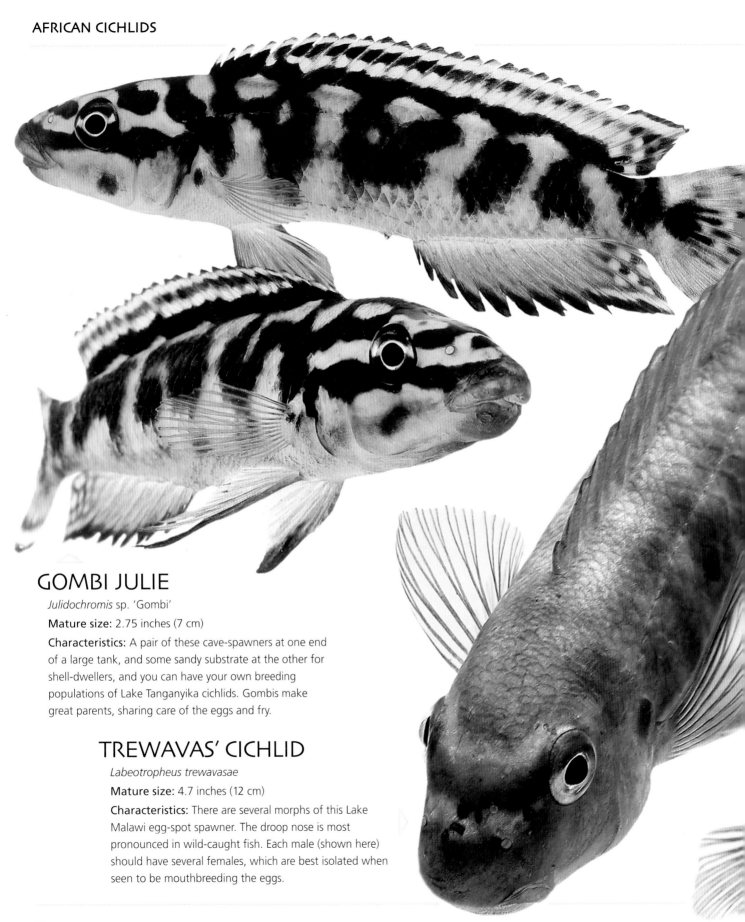

GOMBI JULIE

Julidochromis sp. 'Gombi'

Mature size: 2.75 inches (7 cm)

Characteristics: A pair of these cave-spawners at one end of a large tank, and some sandy substrate at the other for shell-dwellers, and you can have your own breeding populations of Lake Tanganyika cichlids. Gombis make great parents, sharing care of the eggs and fry.

TREWAVAS' CICHLID

Labeotropheus trewavasae

Mature size: 4.7 inches (12 cm)

Characteristics: There are several morphs of this Lake Malawi egg-spot spawner. The droop nose is most pronounced in wild-caught fish. Each male (shown here) should have several females, which are best isolated when seen to be mouthbreeding the eggs.

GOBY CICHLID

Eretmodus cyanosticus

Mature size: 3.2 inches (8 cm)

Characteristics: Living in rock rubble in the shallows of Lake Tanganyika, these fish have a much reduced swim bladder and seldom come off the bottom. Males and females share in mouthbreeding the eggs. They look comic and inoffensive, which sums up their character rather well.

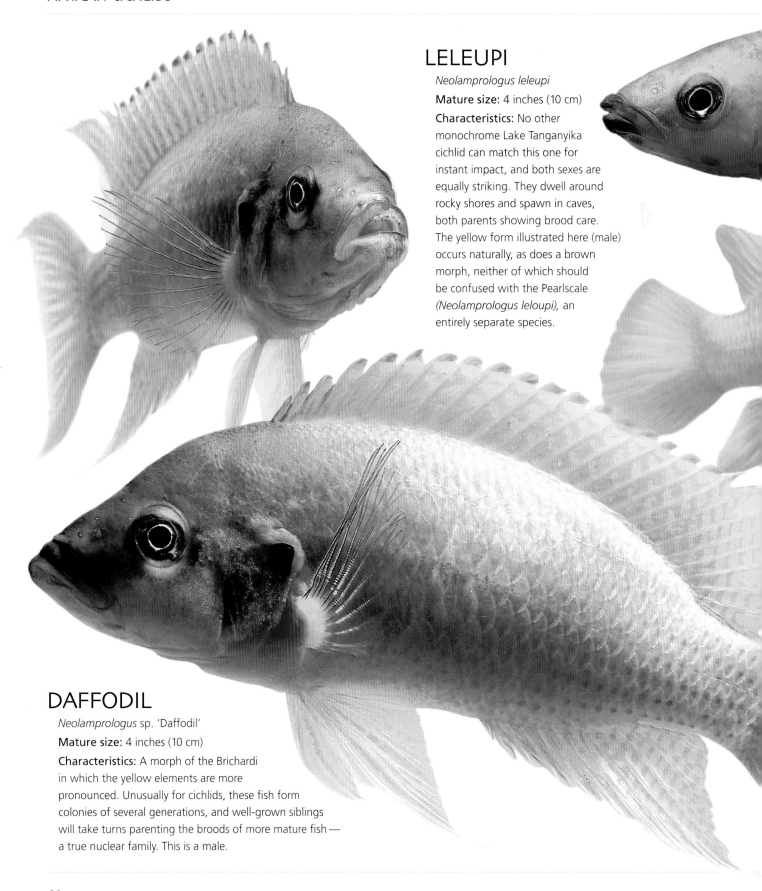

LELEUPI

Neolamprologus leleupi

Mature size: 4 inches (10 cm)

Characteristics: No other monochrome Lake Tanganyika cichlid can match this one for instant impact, and both sexes are equally striking. They dwell around rocky shores and spawn in caves, both parents showing brood care. The yellow form illustrated here (male) occurs naturally, as does a brown morph, neither of which should be confused with the Pearlscale (*Neolamprologus leloupi*), an entirely separate species.

DAFFODIL

Neolamprologus sp. 'Daffodil'

Mature size: 4 inches (10 cm)

Characteristics: A morph of the Brichardi in which the yellow elements are more pronounced. Unusually for cichlids, these fish form colonies of several generations, and well-grown siblings will take turns parenting the broods of more mature fish — a true nuclear family. This is a male.

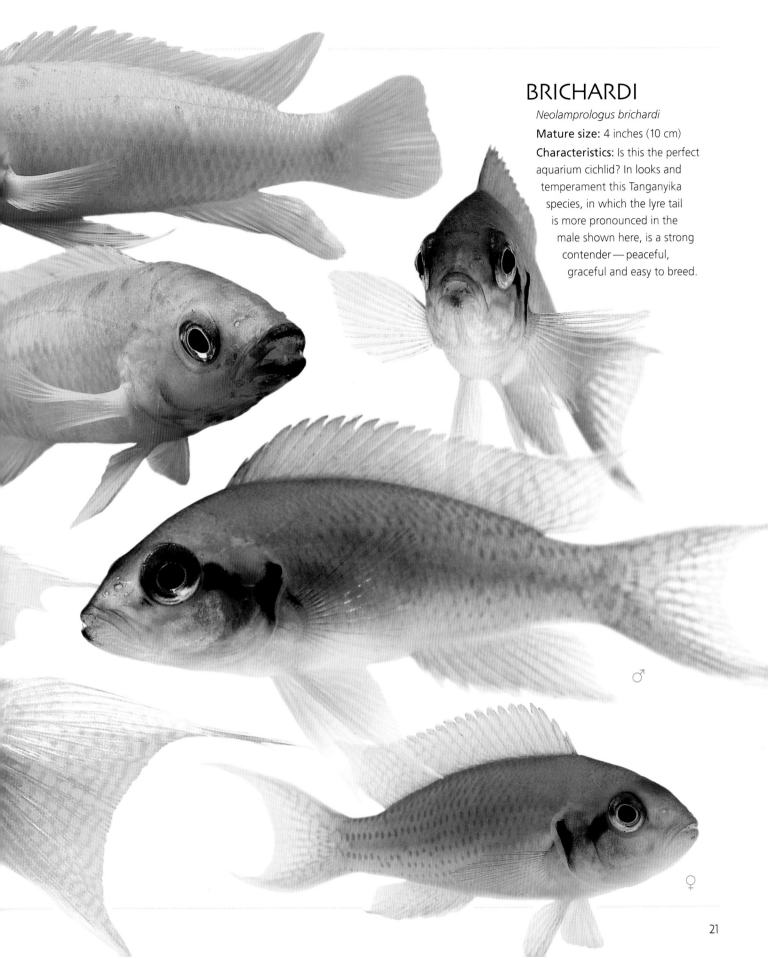

BRICHARDI
Neolamprologus brichardi

Mature size: 4 inches (10 cm)

Characteristics: Is this the perfect aquarium cichlid? In looks and temperament this Tanganyika species, in which the lyre tail is more pronounced in the male shown here, is a strong contender — peaceful, graceful and easy to breed.

♂

♀

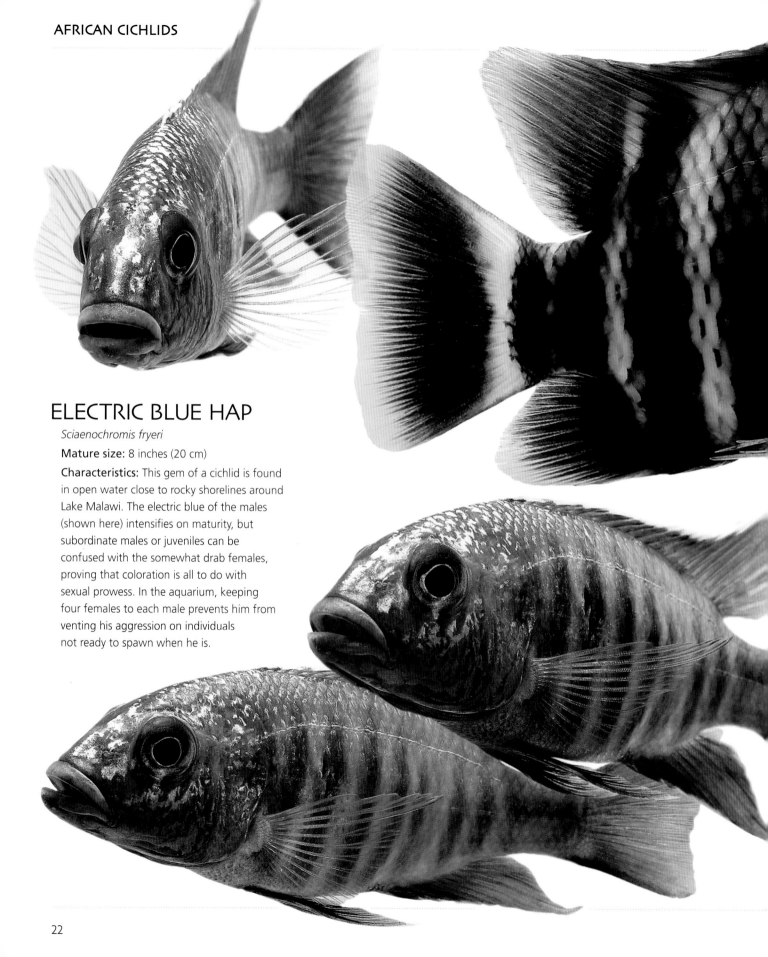

ELECTRIC BLUE HAP

Sciaenochromis fryeri

Mature size: 8 inches (20 cm)

Characteristics: This gem of a cichlid is found in open water close to rocky shorelines around Lake Malawi. The electric blue of the males (shown here) intensifies on maturity, but subordinate males or juveniles can be confused with the somewhat drab females, proving that coloration is all to do with sexual prowess. In the aquarium, keeping four females to each male prevents him from venting his aggression on individuals not ready to spawn when he is.

HORNET TILAPIA

Tilapia buttikoferi

Mature size: 10 inches (25 cm)

Characteristics: A banded beauty from West African rivers and streams. House a compatible pair in a large tank furnished with flat rocks and driftwood for spawning. Both parents look after the free-swimming fry. It may be difficult to move these youngsters on — the Hornet Tilapia has an aggressive reputation — so perhaps this fish is best kept as a singleton, indulgently fed a mixture of earthworms, mosquito larvae and algae wafers.

DUBOISI

Tropheus duboisi

Mature size: 4.7 inches (12 cm)

Characteristics: "Cute" sums up the polka-dotted appeal of young *Tropheus duboisi.* Who could resist them? Sadly, they have to grow up, and in this instance, the adult coloration does not measure up to early promise; the spots soon fade, leaving a rather drab blue individual with a distinctive white transverse belt down each flank. Nonetheless, these dwellers of the deeper rocky coastal zones of Lake Tanganyika are appealing because they are generally not aggressive toward other cichlids. They are maternal mouthbreeders, and although they have been successfully spawned in aquariums, broods are rather small, typically only 5 to 15 eggs.

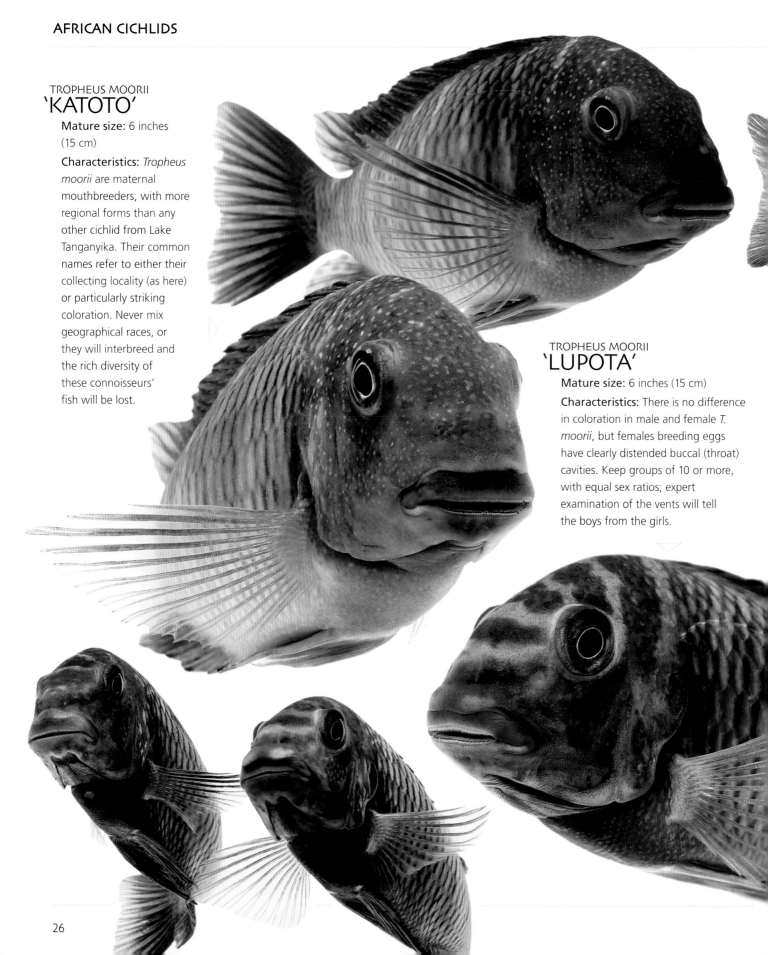

TROPHEUS MOORII
'KATOTO'

Mature size: 6 inches (15 cm)

Characteristics: *Tropheus moorii* are maternal mouthbreeders, with more regional forms than any other cichlid from Lake Tanganyika. Their common names refer to either their collecting locality (as here) or particularly striking coloration. Never mix geographical races, or they will interbreed and the rich diversity of these connoisseurs' fish will be lost.

TROPHEUS MOORII
'LUPOTA'

Mature size: 6 inches (15 cm)

Characteristics: There is no difference in coloration in male and female *T. moorii*, but females breeding eggs have clearly distended buccal (throat) cavities. Keep groups of 10 or more, with equal sex ratios; expert examination of the vents will tell the boys from the girls.

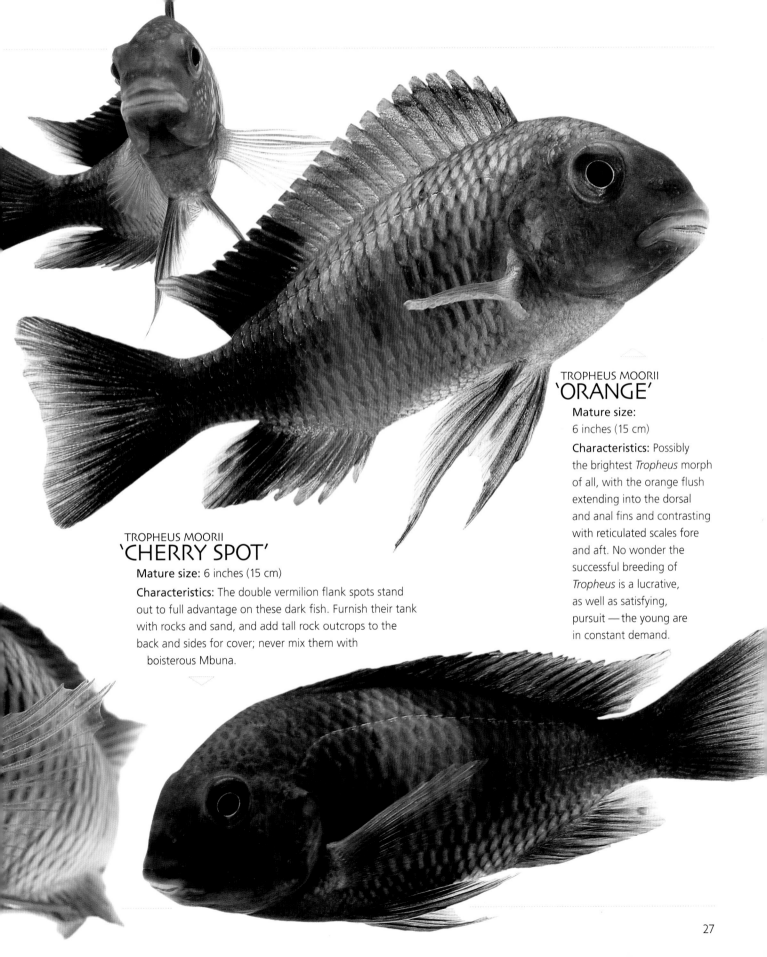

TROPHEUS MOORII
'ORANGE'
Mature size:
6 inches (15 cm)

Characteristics: Possibly the brightest *Tropheus* morph of all, with the orange flush extending into the dorsal and anal fins and contrasting with reticulated scales fore and aft. No wonder the successful breeding of *Tropheus* is a lucrative, as well as satisfying, pursuit — the young are in constant demand.

TROPHEUS MOORII
'CHERRY SPOT'
Mature size: 6 inches (15 cm)

Characteristics: The double vermilion flank spots stand out to full advantage on these dark fish. Furnish their tank with rocks and sand, and add tall rock outcrops to the back and sides for cover; never mix them with boisterous Mbuna.

ICE-BLUE ZEBRA

Metriaclima greshakei

Mature size: 4 inches (10 cm)

Characteristics: Like many Mbuna, this fish from the southeastern arm of Lake Malawi used to be in the genus *Pseudotropheus*, and you will still probably see it described as such in aquatic stores. This is a mature male; females are a uniform rusty brown. Always buy wild-caught fish or named tank-breds; never fish from a dealer's lot of "mixed Malawis." It is the gene pool that will be mixed in all probability, as Mbuna will hybridize.

RUSTY CICHLID

Iodotropheus sprengerae

Mature size: 4 inches (10 cm)

Characteristics: Pair-bonding plays no part in the life of Mbuna. Males (shown here) try to spawn with any females in the vicinity, harassing them if they are not ready. A crowded rockwork aquarium, with crevices in which potential "battered wives" can take refuge, is a good solution to all that testosterone-fueled aggression.

South American cichlids encompass everything from Amazonian dwarfs to the stately Angelfish and the endearing Oscar. Central Americans are beautiful, but can be problematic.

Many cichlid keepers will have cut their teeth on the Convict Cichlid (*Cryptoheros nigrofasciatus*). Long before fashion veered toward the fish of Africa's Rift Lakes, this feisty little Central American was amazing us with its intelligence and devoted brood care. It would spawn anywhere, eat almost anything and put up with indifferent water conditions — but it had another side to its character. When a pair decided to breed, they would claim the entire tank as their own, and woe betide any hapless fish that intruded on their territory.

This is a fact of life with Central American cichlids, most of which are considerably larger than the Convict, and pack correspondingly greater firepower. There are a few peaceable exceptions, such as the cave-spawning *Hypsophrys nicaraguensis* from the lakes of Nicaragua, which will share quarters with the aforementioned Convicts or that other early hobby staple the Firemouth Cichlid (*Thorichthys meeki*). But, as a rule of thumb, it is one pair of "Centrals" per tank — that

Flower Horn Cichlids are man-made hybrids, here shown as a young breeding pair (below) and a vividly patterned adult (right).

Apistogramma viejita typifies cichlid finery in miniature.

would certainly apply to the characterful Midas Cichlid (*Amphilophus citrinellus*) or Jaguar Cichlid (*Parachromis managuensis*). There is even a case for keeping singletons, which — as long as their owners give them plenty of attention — seems to please both parties.

South America offers a much more varied palette of cichlids for

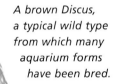

fishkeepers. At one extreme are the dwarfs of the genus *Apistogramma*, rarely growing bigger than 3.2 inches (8 cm). These are polygamous cave-spawners in the wild, but pairs can be successfully bred in soft, neutral to acidic water. Unrelated, but still classed as dwarfs, are cichlids of the genus *Nannacara* and the Ram *(Microgeophagus ramirezi)*. At the other end of the size and temperament scale are Oscars *(Astronotus ocellatus)* and the larger Pike Cichlids of the genus *Crenicichla* — beautiful but challenging.

Occupying the middle ground are Angels *(Pterophyllum scalare* and *P. altum)* — often not regarded as cichlids at all — and longstanding popular subjects such as the Festive Cichlid *(Mesonauta insignis)*, Blue Acara *("Aequidens" pulcher)* and

Severum *(Heros efasciatus)*. Amazonia is also home to the ultimate cichlid challenge, the Discus. There are two wild species, *Symphysodon aequifasciatus* and *S. discus*, but many man-made strains have been

A brown Discus, a typical wild type from which many aquarium forms have been bred.

Young Discus feeding on the nutritious mucus secreted by both their parents.

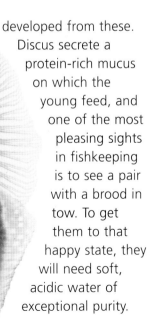

developed from these. Discus secrete a protein-rich mucus on which the young feed, and one of the most pleasing sights in fishkeeping is to see a pair with a brood in tow. To get them to that happy state, they will need soft, acidic water of exceptional purity.

MIDAS CICHLID

Amphilophus citrinellus

Mature size: 12 inches (30 cm)

Characteristics: Everything King Midas touched turned to gold, but things did not work out quite the way he had imagined. If a breeding pair of these characterful cichlids discovers the golden touch, you could be placed in a similar bountiful dilemma, because each spawning can produce 1,000 or more fry. This is a male, with a fatty growth on the forehead called a nuchal hump, which signifies that he is sexually mature.

OSCAR

Astronotus ocellatus

Mature size: 12 inches (30 cm)

Characteristics: An Amazonian cichlid with personality and attitude, that soon comes to recognize its owner. The Oscar is a source of both delight and despair; cute, entreating babies (shown at top) grow rapidly into rapacious, messy feeders that are not to be trusted with smaller fish. Yet their appeal is universal. There are several forms, including albino, red and tiger. All make great pets, and compatible mated pairs are exemplary parents. Do not feed these fish exclusively on worms. They like them too much and become "hooked," refusing anything else.

Tiger

Adult Tiger

Red

LEPIDOTA
PIKE CICHLID

Crenicichla lepidota

Mature size: Males 6.3 inches (16 cm)

Characteristics: They look and feed like the predatory pike, and these Brazilian natives require a spacious, well-filtered tank. So, when buying Pike Cichlids, double-check the species being offered. Dwarf Pikes (from 2.75 inches/7 cm) are a more manageable alternative.

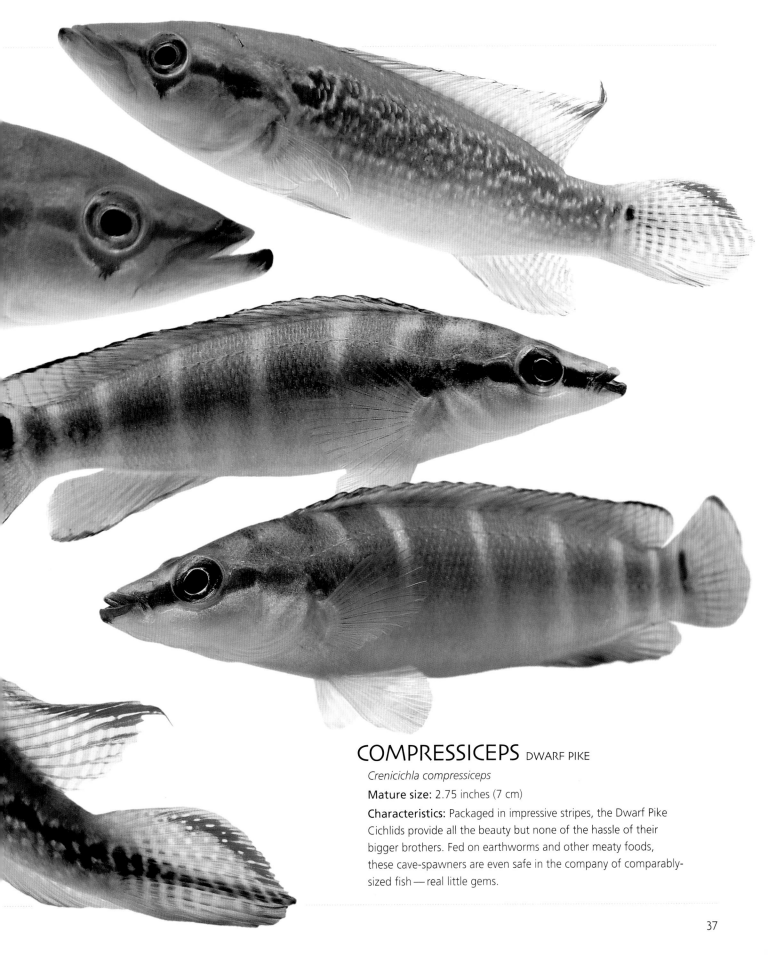

COMPRESSICEPS DWARF PIKE

Crenicichla compressiceps

Mature size: 2.75 inches (7 cm)

Characteristics: Packaged in impressive stripes, the Dwarf Pike Cichlids provide all the beauty but none of the hassle of their bigger brothers. Fed on earthworms and other meaty foods, these cave-spawners are even safe in the company of comparably-sized fish — real little gems.

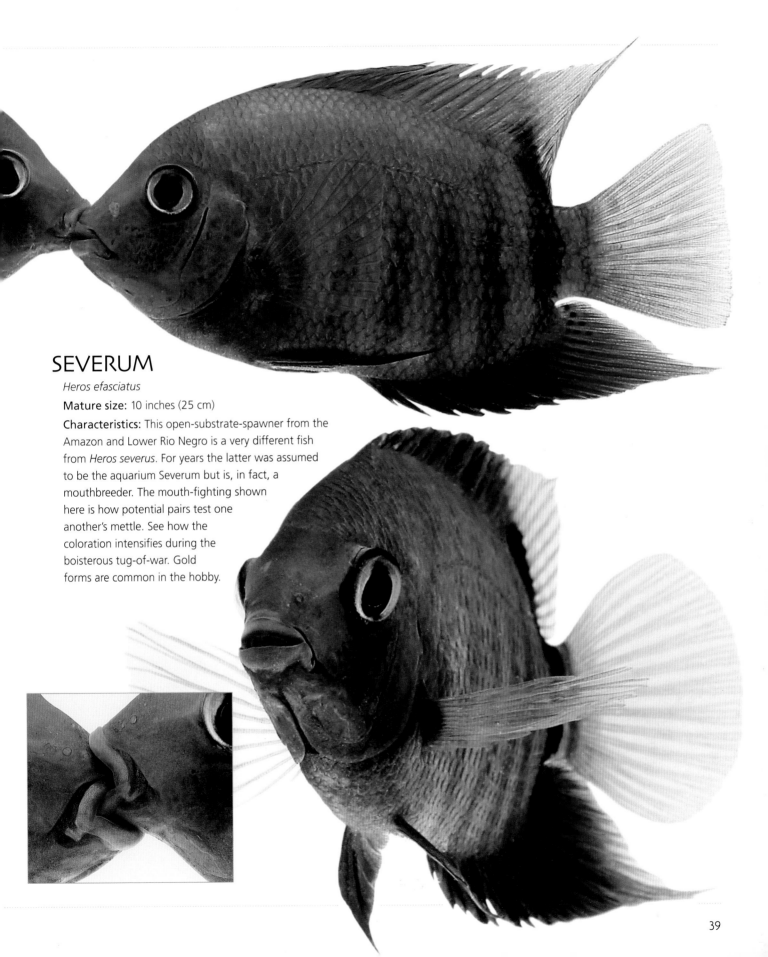

SEVERUM

Heros efasciatus

Mature size: 10 inches (25 cm)

Characteristics: This open-substrate-spawner from the Amazon and Lower Rio Negro is a very different fish from *Heros severus*. For years the latter was assumed to be the aquarium Severum but is, in fact, a mouthbreeder. The mouth-fighting shown here is how potential pairs test one another's mettle. See how the coloration intensifies during the boisterous tug-of-war. Gold forms are common in the hobby.

JAGUAR CICHLID

Parachromis managuensis

Mature size: 12 inches (30 cm)

Characteristics: Many a young "Mannie" has been bought on impulse after performing an entreating dance like this to an aquarium store visitor. These big Central American fish are fully aware of the world outside their tank and, given attention, will strike up a one-to-one relationship with their owners, much like a dog. Their lifespan is on a par with our canine companions too, so a Jaguar Cichlid is a pleasurable long-term commitment.

41

RAM

Microgeophagus ramirezi

Mature size: 2.75 inches (7 cm)

Characteristics: Mass-produced Rams are pallid and unhealthy—sometimes feeding to improve coloration gives a false impression of quality that quickly fades. This male (right) and the pair below are probably wild-caught or hobbyist tank-breds, the safest route to obtaining good specimens.

♀

♂

BLUE-POINT ACARA

"Aequidens" coeruleopunctatus

Mature size: 6 inches (15 cm)

Characteristics: A rarely seen close cousin to the Blue Acara, this cave-spawning cichlid is found in clear, rocky streams from Costa Rica through Panama to Colombia. The male is the lower fish, showing the more pointed anal fin—not an entirely reliable way of sexing them, however, as older females can also develop these fin extensions.

♀

♂

JACK DEMPSEY

"Cichlasoma" octofasciatum

Mature size: 10 inches (25 cm)

Characteristics: This old bruiser from Mexico, Guatemala and Honduras is named after a former world heavyweight boxing champion, yet is no more pugnacious than most other Central American cichlids; with its jutting jaw and stocky build, it just looks that way. These wild-caught examples have clearly had the best attention from their ringside "second"—a fishkeeper who has fallen under their rugged spell. In the pair shown at right, the breeding female is darker than typical aquarium strains.

♀

♂

ALTUM ANGELFISH

Pterophyllum altum

Mature size: 12–15 inches (30–38 cm) tall; 10 inches (25 cm) total length

Characteristics: Much deeper than they are long, these fish from the Orinoco have a sharper gradient from head to jawline than ordinary Angels. The barred flanks blend cryptically into the underwater habitat of dead wood lit by dappled forest light. These rare fish are prized in the hobby; if you can get them to spawn (in soft, acidic water) you are onto a winner!

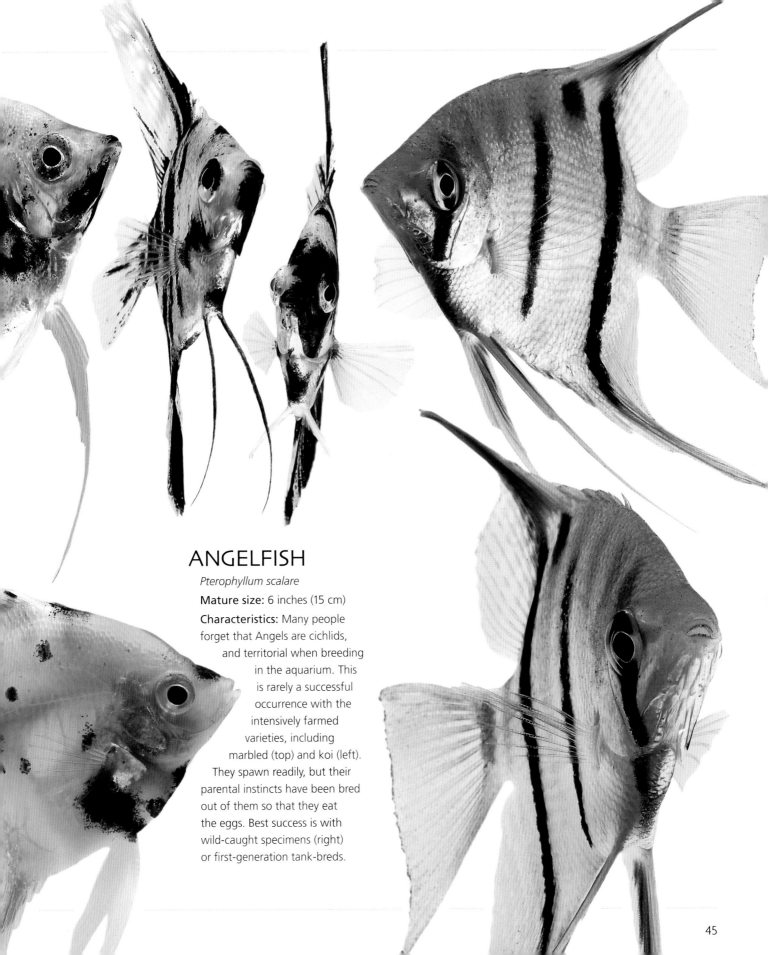

ANGELFISH

Pterophyllum scalare

Mature size: 6 inches (15 cm)

Characteristics: Many people forget that Angels are cichlids, and territorial when breeding in the aquarium. This is rarely a successful occurrence with the intensively farmed varieties, including marbled (top) and koi (left). They spawn readily, but their parental instincts have been bred out of them so that they eat the eggs. Best success is with wild-caught specimens (right) or first-generation tank-breds.

RED MARLBORO DISCUS

Symphysodon aequifasciatus

Mature size: 6 inches (15 cm)

Characteristics: Green and brown Discus, from the Amazon, have been selectively bred into many strains. To spawn, they need deep tanks and soft, pure water. Luck plays its part, too — but what an achievement to see babies feeding off their parents' mucus secretions.

PEARL DISCUS

*Symphysodon
aequifasciatus*

This fish hangs in
the tank like an
opalescent full moon,
cool and mysterious.

PIGEON
BLOOD
DISCUS

*Symphysodon
aequifasciatus*

The term applies to a
perfect ruby, but what
better description of
the scarlet dappling
on this beautiful
Discus?

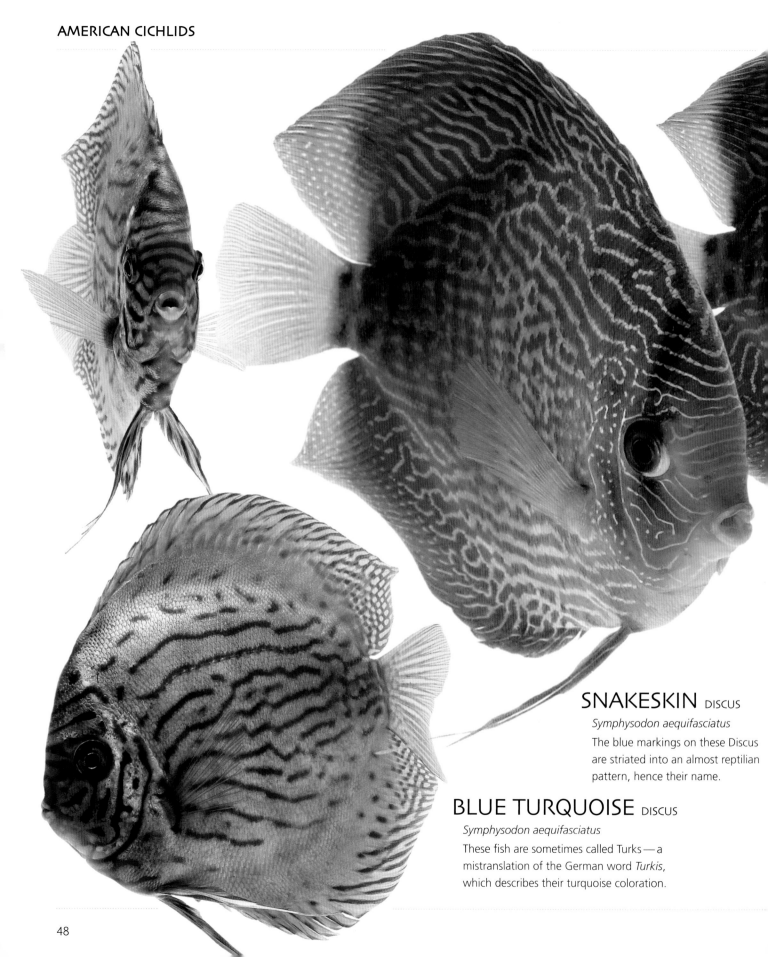

SNAKESKIN DISCUS
Symphysodon aequifasciatus
The blue markings on these Discus are striated into an almost reptilian pattern, hence their name.

BLUE TURQUOISE DISCUS
Symphysodon aequifasciatus
These fish are sometimes called Turks — a mistranslation of the German word *Turkis*, which describes their turquoise coloration.

BLUE DIAMOND DISCUS

Symphysodon aequifasciatus

Breeders of this strain are aiming for a single-hued Discus, and they are almost there. Striking, rather than subtle, the adults have a rich shade yet to be fully achieved by the five-month-old youngsters at right. They are of a size to sell on, but keep small fish such as these in groups of at least 10 to a dozen. Uniform growth in a brood, with no runts, is a mark of good fishkeeping.

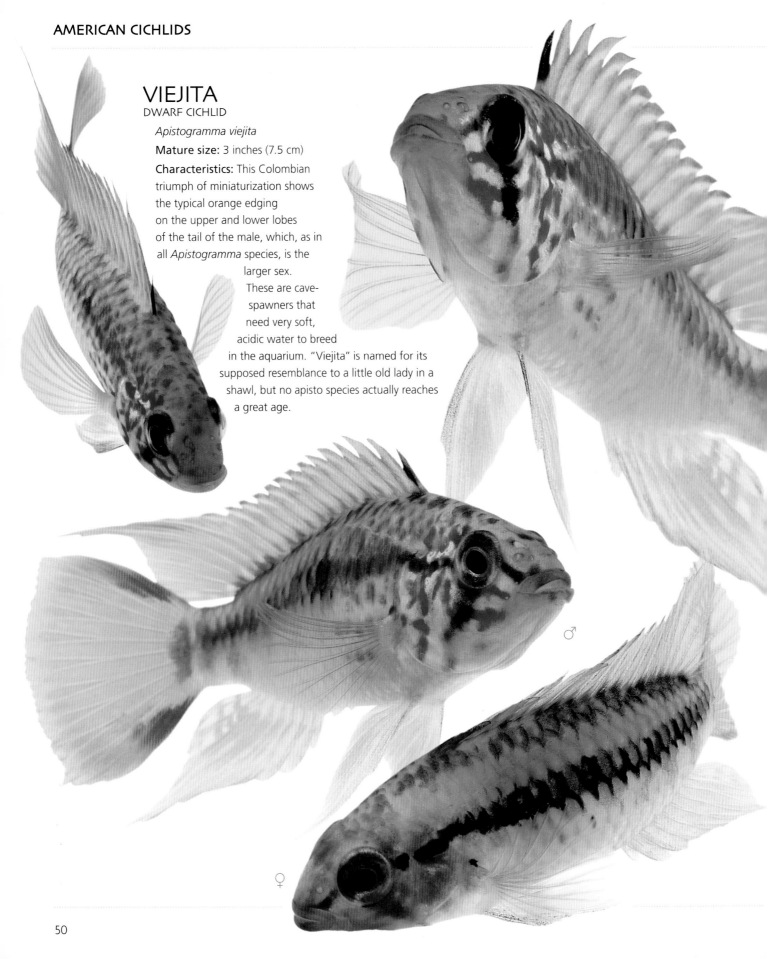

VIEJITA
DWARF CICHLID

Apistogramma viejita

Mature size: 3 inches (7.5 cm)

Characteristics: This Colombian triumph of miniaturization shows the typical orange edging on the upper and lower lobes of the tail of the male, which, as in all *Apistogramma* species, is the larger sex.

These are cave-spawners that need very soft, acidic water to breed in the aquarium. "Viejita" is named for its supposed resemblance to a little old lady in a shawl, but no apisto species actually reaches a great age.

ORANGE AGGIE

Apistogramma agassizii

Mature size: 3.2 inches (8 cm)

Characteristics: A lancet-shaped tail is a common feature of the agassizii group. This fish is a cultivated orange form, with the typical caudal fin shape, but far removed from anything you would encounter in this dwarf cichlid's native Amazonian habitat of streams and pools.

COCKATOO
DWARF CICHLID

Apistogramma cacatuoides

Mature size: 2 inches (5 cm)

Characteristics: This "double red" form has helped to boost the popularity of Cockatoo Dwarf Cichlids in the hobby. In wild fish from the Amazon Basin, the tails are nowhere near as bright, and those of the females are rounded. In "double reds," females' tails can be slightly lyrate too, which makes sexing the young difficult.

You could fill an aquarium with nothing but characins and still enjoy scenes of endless interest. They include universally popular fish, as well as some for the specialist.

Here we are really looking at two groups of fish spread across Africa and South America. There are tetras — small, peaceable shoalers, often farm-bred and highly adaptable to aquarium life — plus a diverse selection known as characins. In other words, all tetras are characins, but not all characins are tetras.

Tetras are large-eyed and keen-sighted. Mainly from the Amazon catchment, they seem lit from

*A Red Hook (*Myleus rubripinnis*), one of the "Silver Dollar" group.*

within as their glittering scales refract the light in all shades of the spectrum. Most are egg-scatterers and show no brood care. They feed on small aquatic insects and crustaceans.

Tetra genera yielding the richest vein of aquarium jewels are *Hyphessobrycon* (Bleeding Heart, Black Phantom, Lemon, Serpae and Flag Tetras) and *Hemigrammus* (Rummy-nose, Head-and-tail-light,

Glowlight, Golden and Featherfin Tetras). But a very small group deservedly tops the popularity stakes on both sides of the Atlantic: the Neon Tetra (*Paracheirodon innesi*) and the even more vivid Cardinal (*P. axelrodi*). These fish cost a week's salary

Leporinus species, such as this L. fasciatus, *have mouths adapted to algae-grazing.*

CINS

The Glowlight Tetra reflects light from a stripe of iridescent scales along the head and body.

when they first appeared in the hobby, but now, thanks to intensive farming, they can be had for pocket change. The same is true of most tetras. Wild-caught specimens cannot be beaten for bright markings, but are harder to acclimatize to aquarium life and require soft, acidic water.

From beauty to the beast — the Red Belly Piranha (Serrasalmus nattereri) is commonly kept, despite its fearsome reputation. There are also several convincing lookalikes, namely "Silver Dollars"

Ever-popular Cardinal (above) and Neon Tetras.

Hatchetfish mirrored at the surface, their "pectoral wings" ready to fly.

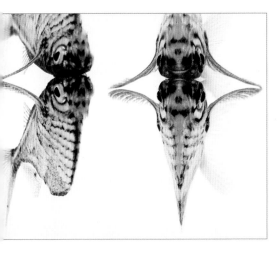

of the genera Metynnis, Myleus and Mylossoma, which eat nothing more controversial than fruit and soft-leaved plants.

Still in South America, tiny, surface-swimming hatchetfish are ideal for community tanks. They have deep bodies and use their pectoral fins as wings to leap into the air to escape predators.

Pencilfish are peaceful, slimline beauties that spend much of their time in midwater, often in a head-up position. One of them, Nannostomus eques, has a bold lateral stripe extending into the lower tail lobe, explaining its common name of Hockey Stick

Tetra. This feature is shared by the Penguin Tetra (Thayeria boehlkei).

For bold patterning, it is hard to beat the Striped Headstander (Anostomus anostomus), whose markings run from head to tail, or the Banded Leporinus (Leporinus fasciatus), which seems to be wearing a black-and-yellow sweater.

Relatively few African characins come into the hobby, but the Congo Tetra (Phenacogrammus interruptus) is a noteworthy exception. For the larger aquarium there are two species of the algae-grazing Distichodus species, with vertical barring as sharp and clear as on any Asian Clown Loach.

LONG-NOSED
DISTICHODUS

Distichodus lusosso

Mature size: 16 inches (40 cm)

Characteristics: Most aquarium characins originate from South America, but this barred beauty is from Africa's Zaire Basin. The fatty adipose fin (behind the dorsal) is a characin trademark, although nobody is sure of its use, or even if it has one. In the wild, these fish graze on algae growing on rocks, and a brightly lit aquarium will encourage a supply of this vegetarian fare. Note the adult size — these are not suitable for small or heavily planted tanks, whose decor they will uproot, and they can be aggressive toward smaller fish.

ROUND-FACED
DISTICHODUS

Distichodus sexfasciatus

Mature size: 39 inches (100 cm)

Characteristics: A little Latin helps enormously in understanding your fish; *sexfasciatus* means six-banded. The underslung, hard-lipped mouth of this giant suggests specialized algae-grazing in rock crevices. The scales reach well into the wrist of the deeply forked tail. This species is reputed to be more peaceful than *D. lusosso*, and is an egg-scatterer, but has never been bred in home aquariums. There are no known external sexual differences.

Long-nosed

Round-faced

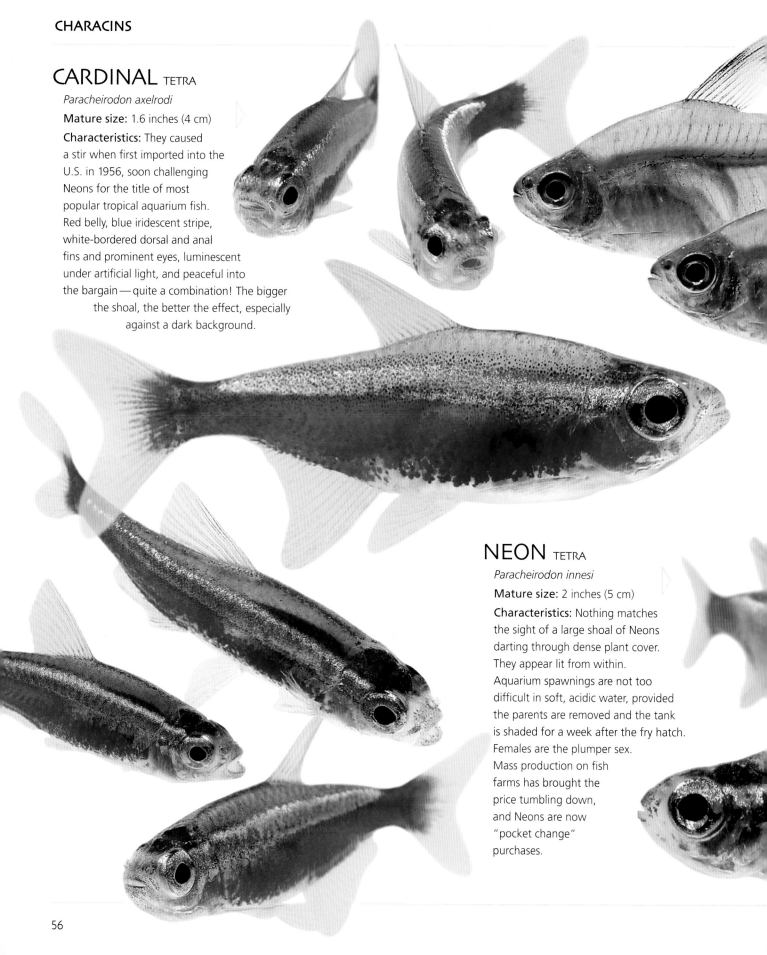

CARDINAL TETRA

Paracheirodon axelrodi

Mature size: 1.6 inches (4 cm)

Characteristics: They caused a stir when first imported into the U.S. in 1956, soon challenging Neons for the title of most popular tropical aquarium fish. Red belly, blue iridescent stripe, white-bordered dorsal and anal fins and prominent eyes, luminescent under artificial light, and peaceful into the bargain—quite a combination! The bigger the shoal, the better the effect, especially against a dark background.

NEON TETRA

Paracheirodon innesi

Mature size: 2 inches (5 cm)

Characteristics: Nothing matches the sight of a large shoal of Neons darting through dense plant cover. They appear lit from within. Aquarium spawnings are not too difficult in soft, acidic water, provided the parents are removed and the tank is shaded for a week after the fry hatch. Females are the plumper sex. Mass production on fish farms has brought the price tumbling down, and Neons are now "pocket change" purchases.

LEMON TETRA

Hyphessobrycon pulchripinnis

Mature size: 1.8 inches (4.5 cm)

Characteristics: A stylish rather than flashy dresser, this translucent little beauty has no vices. Treat yourself to a shoal of six or more, mixing the sexes so that they display their best coloration. The dark leading edge to the anal fin is more pronounced in males, while females have the higher back.

RED BELLY PIRANHA

Serrasalmus nattereri

Mature size: 11 inches (28 cm)

Characteristics: The reputation of this fish goes before it, and the morbid fascination factor is largely responsible for its popularity. The spotted youngsters will certainly nip and fight among themselves, so choosing a perfect specimen can be difficult. In maturity, the Red Belly Piranha is something of an introvert (except at feeding time), and needs a well-furnished tank and the company of shoalmates. Aquarium spawnings are quite common, the fish scattering eggs among plants.

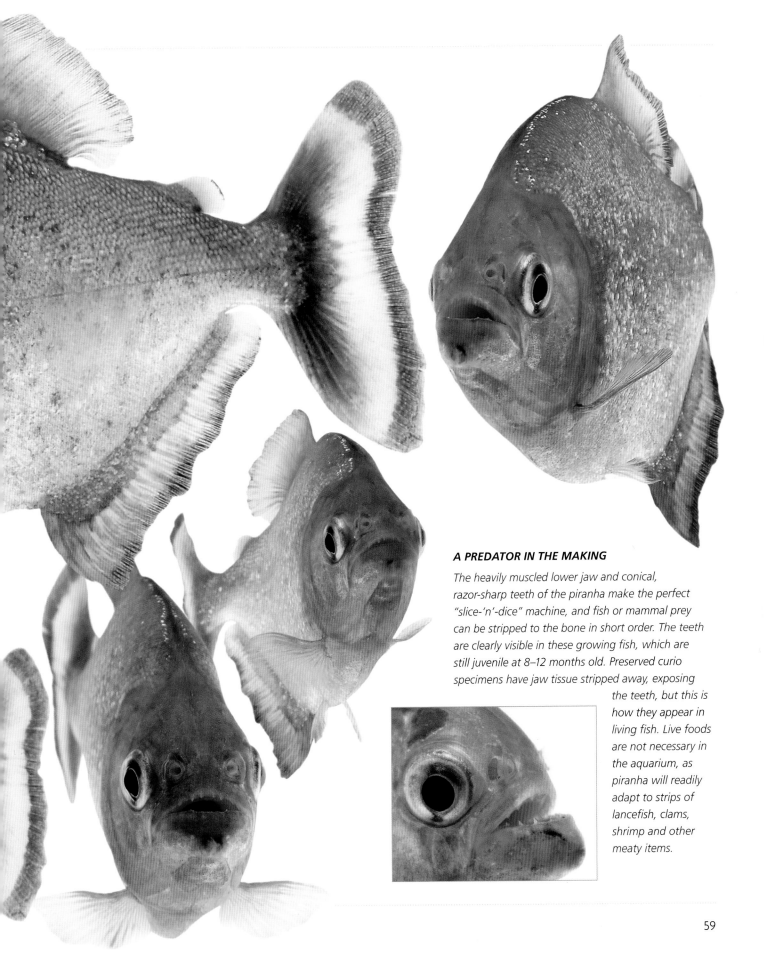

A PREDATOR IN THE MAKING

The heavily muscled lower jaw and conical, razor-sharp teeth of the piranha make the perfect "slice-'n'-dice" machine, and fish or mammal prey can be stripped to the bone in short order. The teeth are clearly visible in these growing fish, which are still juvenile at 8–12 months old. Preserved curio specimens have jaw tissue stripped away, exposing the teeth, but this is how they appear in living fish. Live foods are not necessary in the aquarium, as piranha will readily adapt to strips of lancefish, clams, shrimp and other meaty items.

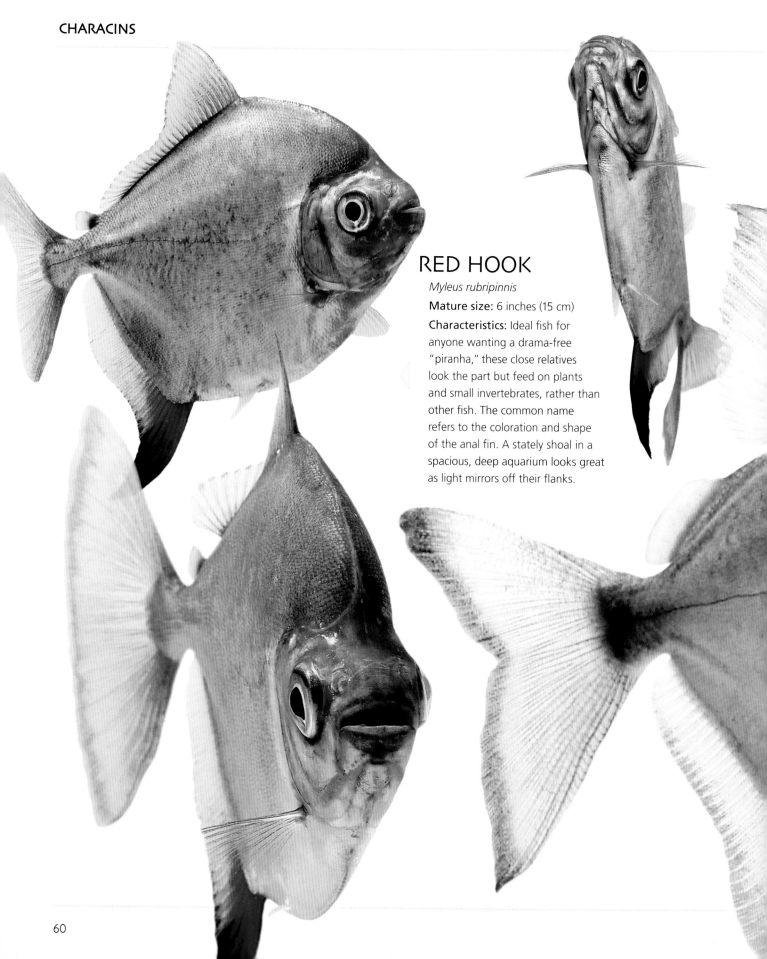

RED HOOK

Myleus rubripinnis

Mature size: 6 inches (15 cm)

Characteristics: Ideal fish for anyone wanting a drama-free "piranha," these close relatives look the part but feed on plants and small invertebrates, rather than other fish. The common name refers to the coloration and shape of the anal fin. A stately shoal in a spacious, deep aquarium looks great as light mirrors off their flanks.

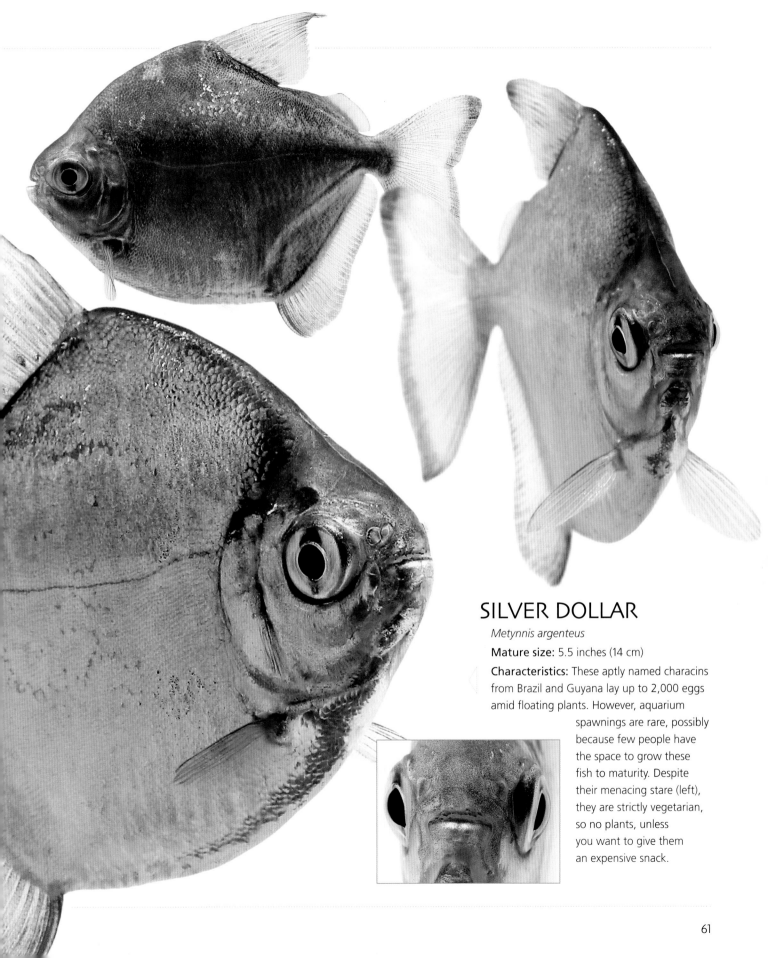

SILVER DOLLAR

Metynnis argenteus

Mature size: 5.5 inches (14 cm)

Characteristics: These aptly named characins from Brazil and Guyana lay up to 2,000 eggs amid floating plants. However, aquarium spawnings are rare, possibly because few people have the space to grow these fish to maturity. Despite their menacing stare (left), they are strictly vegetarian, so no plants, unless you want to give them an expensive snack.

BLEEDING HEART TETRA

Hyphessobrycon erythrostigma

Mature size: 2.4 inches (6 cm)

Characteristics: The red flank patch gives this Upper Amazon fish its common name. Males have dorsal and anal fin extensions, making sexing of mature specimens easy. Aquarium stalwarts and deservedly popular.

♂

♀

GLOWLIGHT TETRA

Hemigrammus erythrozonus

Mature size: 1.6 inches (4 cm)

Characteristics: These tiny fish appear to glow, leading newcomers to believe that they generate their own light. A convincing illusion, probably assisting shoals to stay close-knit in their sun-dappled Guyanan home waters.

CONGO TETRA

Phenacogrammus interruptus

Mature size: 3.2 inches (8 cm)

Characteristics: The males (here) sport elongated dorsal and caudal finnage, and both sexes have large, reflective scales that show best against a dark background and substrate. These active swimmers from Zaire require a large tank.

SERPAE TETRA

Hyphessobrycon eques

Mature size: 1.6 inches (4 cm)

Characteristics: Serpae Tetras have been a popular choice for aquariums for many years, and today these deep-bodied fish from the Southern Amazon are commercially bred by the million. The black dorsal fin and flank flashes contrast well with the rich, coppery body through which the swim bladder can be seen to shine.

PENGUIN TETRA

Thayeria boehlkei

Mature size: 2.4 inches (6 cm)

Characteristics: Hockey Stick is another name for these dapper little fish, which can produce up to 1,000 eggs at every spawning. Females are deeper in the body than males, and both sexes tend to swim with the head slightly raised.

MARBLED HATCHETFISH

Carnegiella strigata

Mature size:
1.6 inches (4 cm)

Characteristics: These deep-bodied characins, lacking an adipose fin, are surface-dwellers that appreciate a strong current and pure, soft water. Eggs can just be seen developing in the body of the female (top), while males are more strikingly marked.

♂

♀

BANDED
LEPORINUS

Leporinus fasciatus

Mature size: 10 inches (25 cm)

Characteristics: Do not give this
denizen of the Amazon tributaries a planted
home, as it will dine on the decor. Although
also called Headstander, it usually swims on
an even keel, unlike another "headstander,"
the horizontally barred *Anostomus ternetzi*.
These fish prefer a cool tank.

Giving birth to live young is a much more efficient means of reproduction than either random egg-scattering or parental brood care, since the fish come into the world fully formed and ready to fend for themselves.

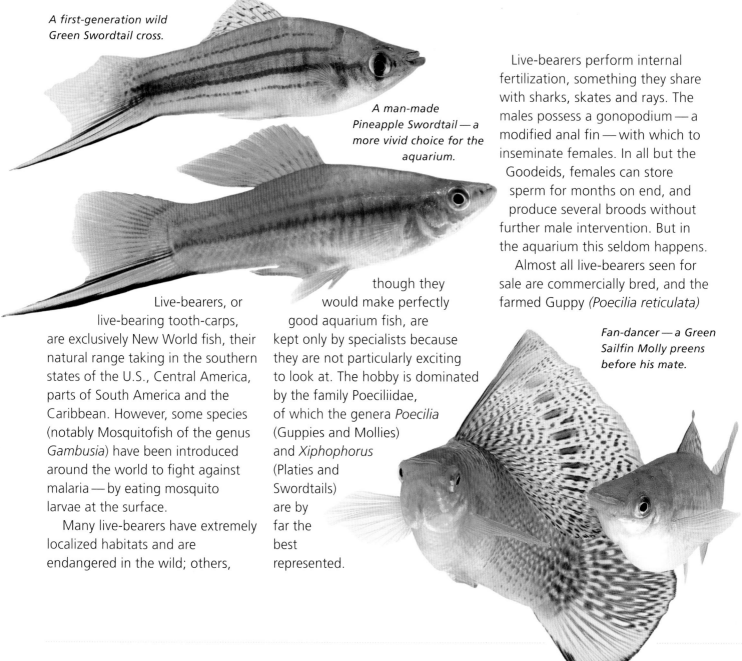

A first-generation wild Green Swordtail cross.

A man-made Pineapple Swordtail — a more vivid choice for the aquarium.

Fan-dancer — a Green Sailfin Molly preens before his mate.

Live-bearers perform internal fertilization, something they share with sharks, skates and rays. The males possess a gonopodium — a modified anal fin — with which to inseminate females. In all but the Goodeids, females can store sperm for months on end, and produce several broods without further male intervention. But in the aquarium this seldom happens.

Almost all live-bearers seen for sale are commercially bred, and the farmed Guppy (*Poecilia reticulata*)

Live-bearers, or live-bearing tooth-carps, are exclusively New World fish, their natural range taking in the southern states of the U.S., Central America, parts of South America and the Caribbean. However, some species (notably Mosquitofish of the genus *Gambusia*) have been introduced around the world to fight against malaria — by eating mosquito larvae at the surface.

Many live-bearers have extremely localized habitats and are endangered in the wild; others, though they would make perfectly good aquarium fish, are kept only by specialists because they are not particularly exciting to look at. The hobby is dominated by the family Poeciliidae, of which the genera *Poecilia* (Guppies and Mollies) and *Xiphophorus* (Platies and Swordtails) are by far the best represented.

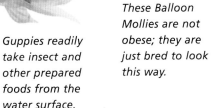

is a world away from the wild Mexican population. Selective breeding continues to "improve" on nature, with wildly extravagant tail and dorsal finnage and every imaginable coloration readily available. However, it is hard to improve upon the wild Green Sailfin Molly (Poecilia velifera), from Mexico and Yucatan — all breeders have done is to cultivate the albino, or golden, form.

Platies (Xiphophorus maculatus) are produced in high-fin varieties, but most retain their original sturdy body shape and short finnage, and it is only the coloration that has been experimented with. Commercial 'Variatus' platies probably carry Swordtail blood, as crosses with Xiphophorus helleri are quite viable.

To breed Platies, Guppies or Swords true to their ornamental strain, you need fish of proven lineage

These Balloon Mollies are not obese; they are just bred to look this way.

Guppies readily take insect and other prepared foods from the water surface.

and "virgin" females; this can be difficult, as they can be sexually mature within a matter of weeks. Most fishkeepers are happy just to see successive broods produced and a self-perpetuating live-bearer population of "mongrels."

NEON BLUE
GUPPY

Poecilia reticulata

Mature size: 2.4 inches (6 cm)

Characteristics: The fish featured here are all delta-tailed male guppies. The various strains, bearing exotic trade names such as Neon Blue, Snakeskin, Green Lace, Sunrise, King Cobra, Half-black or Disco, may breed true if kept apart. But most hobbyists are content with a mixture; seeing how the young turn out only adds to the excitement. Pregnant females show a dark gravid spot on the abdomen.

KING COBRA GUPPY

Poecilia reticulata

Mature size: 2.4 inches (6 cm)

Characteristics: Unlike their drab wild ancestors, farmed guppies are available in almost limitless permutations of coloration and fin shape. Selective breeding has given the males exaggerated, flowing tails, used in continual courtship displays. The plainer, more rotund females are impregnated by the modified anal fin (gonopodium) of their suitors, and give birth to up to 40 live young.

SUNRISE GUPPY

Poecilia reticulata

Mature size: 2.4 inches (6 cm)

Characteristics: Guppies are originally from Central America, but will be happy in a wide temperature band (64–82°F / 18–28°C), given hard, neutral water. Avoid keeping them with boisterous barbs or male Betta of the same coloration, which may mistake Guppies for rivals and attack them.

GREEN SAILFIN MOLLY

Poecilia velifera

Mature size: 7 inches (18 cm)

Characteristics: Both the wild-caught fish from Mexico and Yucatan and a cross between *Poecilia velifera* and *P. latipinna* bear the same common name. Both need plenty of green food and, given the male's amazing but vulnerable dorsal fin, these fish do best in a species rather than a community tank. They prefer hard water to which some marine salt has been added.

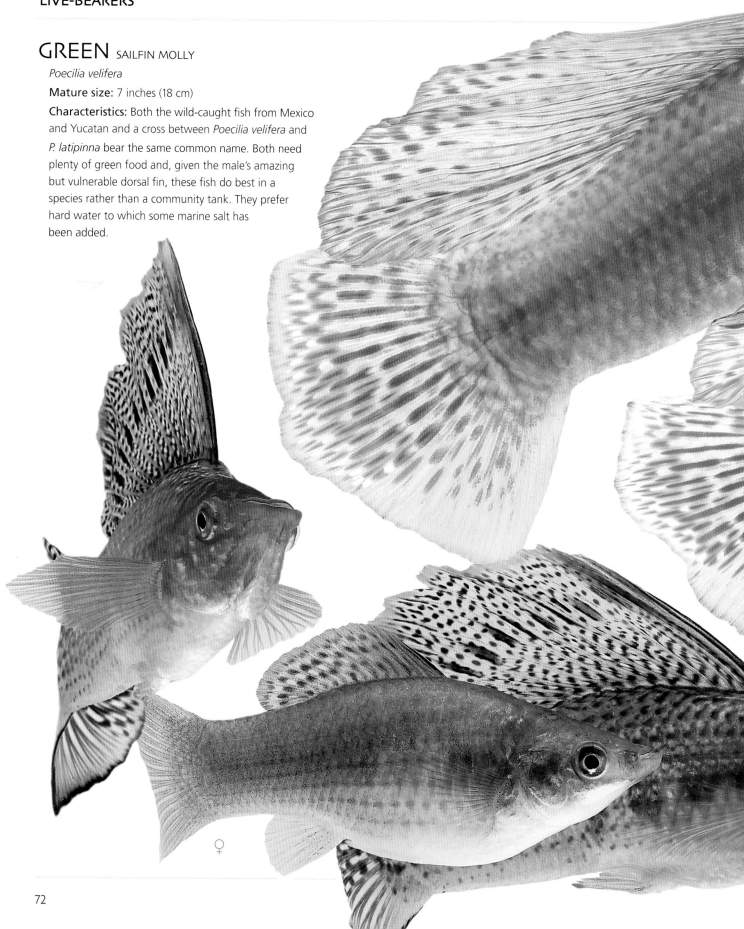

♀

ORANGE SAILFIN MOLLY

Poecilia velifera

Mature size: 7 inches (18 cm)

Characteristics: This farm-bred albino version of the Green Sailfin Molly lacks all the dark pigment. The sailfin characteristics are exaggerated in this fish. Note the gonopodiums in the two males shown below.

SILVER SAILFIN MOLLY

Poecilia spp.

Mature size: 7 inches (18 cm)

Characteristics: This cross between the two most common Molly species shows a lyre tail and a moderately tall dorsal fin. The more extreme Silver Balloon Molly has a deep, foreshortened body.

BLUE ONE-SPOT
PLATY

Xiphophorus maculatus

Mature size: 2.4 inches
(6 cm)

Characteristics: In these live-bearers, there are countless variations of coloration and finnage over the original Mexican ancestor. Sexes are easily told apart by the anal fin; in the male it has become modified into an internal organ of fertilization known as a gonopodium.

Male   *Female*

ORANGE PLATY

Xiphophorus maculatus

Mature size: 2.4 inches (6 cm)

Characteristics: In the best examples of these peaceable fish, the startlingly vivid body coloration is carried through into all the fins. These are males.

RED WAGTAIL PLATY

Xiphophorus maculatus

Mature size: 2.4 inches (6 cm)

Characteristics: With fin rays picked out in black and a uniform crimson body, these live-bearers (here, males) are the ultimate in everyday aquarium chic.

TUXEDO PLATY

Xiphophorus maculatus

Mature size: 2.4 inches (6 cm)

Characteristics: In their smart but understated formal livery, these Platies are a good foil to examples with more showy coloration. Be aware that all strains readily interbreed, so try to keep them separate. These are females.

PINEAPPLE
SWORDTAIL

Xiphophorus helleri

Mature size: 4 inches
(10 cm)

Characteristics:
The wild fish, from
Central America, is an
unexceptional green, but the
breeders have got to work to
produce several striking varieties.
Males have the long, swordlike
extension to the lower lobe of
the tail fin, but this comes only
with time. The females are larger
and more thickset.

BLACK SWORDTAIL

Xiphophorus helleri

Mature size: 4 inches (10 cm)

Characteristics: This is the Swordtail
version of the closely related Tuxedo
Platy, and interbreeding of the two
species is suspected.

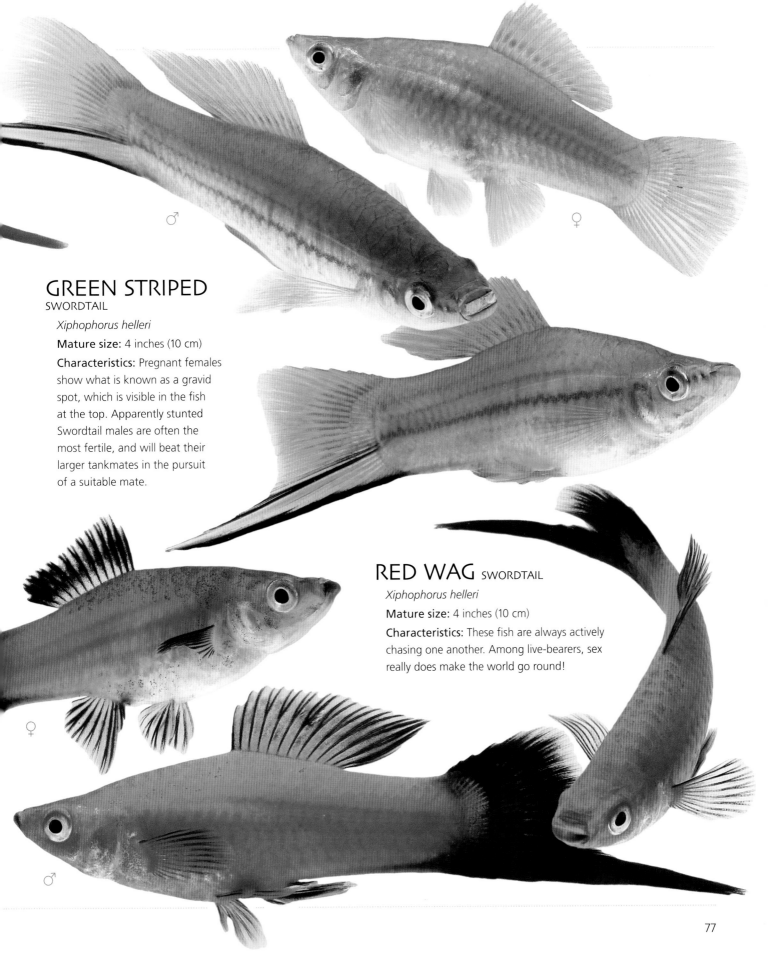

GREEN STRIPED
SWORDTAIL

Xiphophorus helleri

Mature size: 4 inches (10 cm)

Characteristics: Pregnant females show what is known as a gravid spot, which is visible in the fish at the top. Apparently stunted Swordtail males are often the most fertile, and will beat their larger tankmates in the pursuit of a suitable mate.

RED WAG SWORDTAIL

Xiphophorus helleri

Mature size: 4 inches (10 cm)

Characteristics: These fish are always actively chasing one another. Among live-bearers, sex really does make the world go round!

Fish eggs through the mail? That is how some enthusiasts increase their stocks of these incandescent little fish, whose lives flame all too briefly.

KILLI

While killifish are not mainstream in the hobby, they are fascinating to keep, and ideal for anyone without the space to run a full-scale aquarium. Most species can be housed and bred in small tanks. Properly known as egg-laying tooth-carps, they are widely distributed across Africa, Asia and the Americas. Their closest relatives are familiar live-bearers such as Guppies and Platies, but, because of the often hostile nature of their environment, some have developed ingenious breeding strategies to ensure survival through the generations.

To get the best from killifish, you should try to spawn them, if only to maintain stocks. Those from savanna regions, for example, inhabit ponds that dry up every year. Before that happens and the parent fish die, tough eggs are deposited, like tiny time capsules, in the bottom mud. These hatch when the rains return. These killies are known as "annuals" and,

Fundulopanchax gardneri nigerianum — a popular killie among hobbyists.

although they may survive beyond a year in the aquarium, they are by definition short-lived. Yet their beauty is unrivaled, the males with their large dorsal and anal fins, the tail often lyre-shaped, and the whole body suffused with spangled patterns of almost unreal brightness. The West African genus *Fundulopanchax* is the most popular of all.

Once their eggs are laid, collect and store them in damp peat moss in a warm, dark place for the incubation period, which can be several months. Then add water from an established killie tank and the fry will hatch in hours. With

time short to attain maturity, they grow fast and need copious amounts of small, live food items.

Killifish from more stable habitats lay their eggs on the roots of floating plants, but in the hobby a spawning mop is substituted.

FISH

Like other killifish, this male Fundulopanchax walkeri *is more showy than the female.*

Killies are small, but beautifully marked. Tail design varies among males, but all are aimed at attracting mates.

The flowing fins and bold markings of the Cape Lopez Lyretail (Aphyosemion australe).

Eggs can be picked off this or left in place and stored in a jar of tank water, where they will hatch in two to three weeks. Typical plant-spawners are killies of the genus *Epiplatys* (commonly known as Panchax) from West Africa. They look like miniature pike, and need to be kept in a species tank or with larger tankmates, as they are predatory.

One of the best-known killifish, the American Flagfish (*Jordanella floridae*) from Florida, is often not recognized as such, because it looks like a Platy and its reproductive behaviour is more akin to that of a cichlid, with the male caring for the brood. Killies are full of surprises.

GARDNERI FUNDULOPANCHAX

Gardner's Aphyosemion

Mature size: 2.4 inches (6 cm)

Characteristics: This is a yellow variety— others are blue. Males (here) are brighter, while females of the various subspecies all look much alike and are a pale tan. These fish fare best in a shaded species tank and are known as substrate-spawners, laying their eggs in bottom detritus. A ratio of three females to each male brings the best chance of spawning success.

APHYOSEMION OGOENSE
PYROPHORE

Mature size: 2 inches (5 cm)

Characteristics: The candy-striped fins and lyre tail of the male make this a sought-after, though rather rare, West African killie. After a courtship of much chasing and display, adhesive eggs are laid on plants or, in the aquarium, on a spawning mop. Note the large eyes set well up in the head—typical of a fish that is both predator and prey.

♂

♀

FUNDULOPANCHAX GARDNERI
NIGERIANUM

Steel-blue Lyretail

Mature size: 2.4 inches (6 cm)

Characteristics: Their common name does not do these little gems justice — any artists painting them as nature has done would be accused of letting their imagination run riot. Rivaling any cultivated Guppy, these wild fish are from Nigeria.

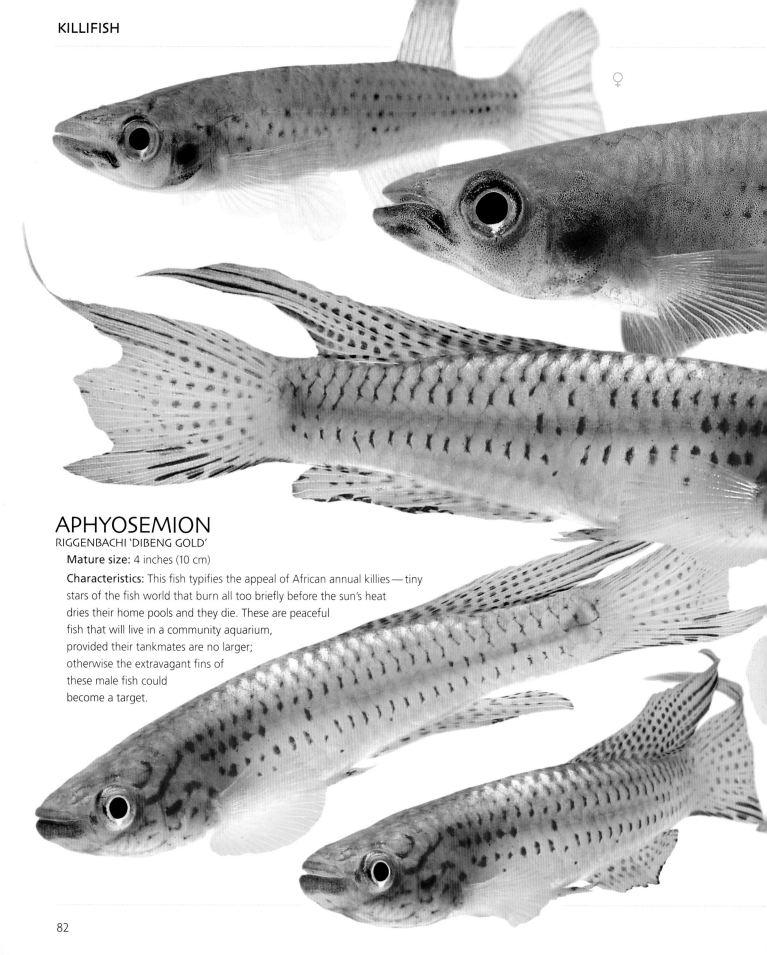

♀

APHYOSEMION
RIGGENBACHI 'DIBENG GOLD'

Mature size: 4 inches (10 cm)

Characteristics: This fish typifies the appeal of African annual killies—tiny stars of the fish world that burn all too briefly before the sun's heat dries their home pools and they die. These are peaceful fish that will live in a community aquarium, provided their tankmates are no larger; otherwise the extravagant fins of these male fish could become a target.

♂

EPIPLATYS
CHAPERI ANGONA

Mature size: 2.5 inches (6.5 cm)

Characteristics: New subspecies of *Epiplatys chaperi* from West Africa continue to come into the hobby. In this pair, the difference between the sexes in terms of size and coloration is clearly illustrated. The glory of the male lies not so much in its subtle body livery as in its large orange pectoral fins, compared to those of the female, which are clear.

EPIPLATYS ROLOFFI

Roloff's Panchax

Mature size: 2.75 inches (7 cm)

Characteristics: Like all Panchax, these West African fish have large mouths suited to taking insect prey from the surface, where they spend much of their time. They lay adhesive eggs on plants and so can be bred on spawning mops. The blue head of the male (below) makes a superb contrast with the reticulated scales and rainbow-edged finnage.

♀

♂

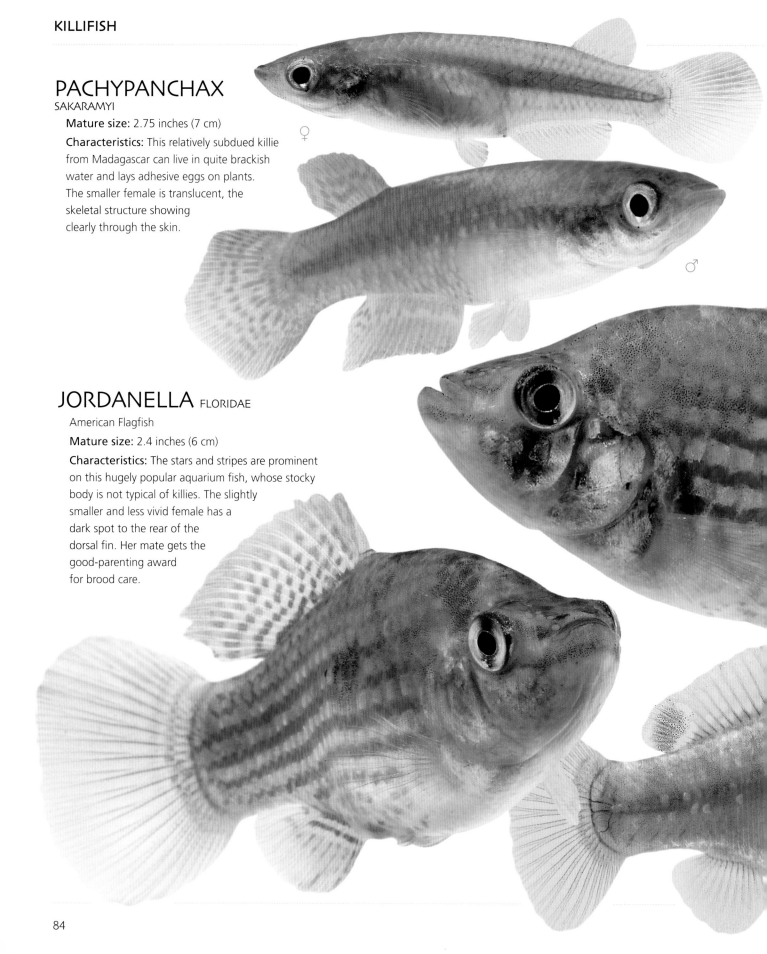

PACHYPANCHAX
SAKARAMYI

Mature size: 2.75 inches (7 cm)

Characteristics: This relatively subdued killie from Madagascar can live in quite brackish water and lays adhesive eggs on plants. The smaller female is translucent, the skeletal structure showing clearly through the skin.

JORDANELLA FLORIDAE

American Flagfish

Mature size: 2.4 inches (6 cm)

Characteristics: The stars and stripes are prominent on this hugely popular aquarium fish, whose stocky body is not typical of killies. The slightly smaller and less vivid female has a dark spot to the rear of the dorsal fin. Her mate gets the good-parenting award for brood care.

FUNDULOPANCHAX
SPOORENBERGI

Spoorenberg's Panchax

Mature size: 4.7 inches (12 cm)

Characteristics: These substrate-divers grow to a good size (for killies). The yellow-edged tails and fins of these males almost glow, and are shown to best effect when they are displaying to potential mates. Killifish prefer soft, neutral water, but will survive in less suitable tank environments too.

We can Scuba dive in a watery world. Anabantids turn the tables, breathing air at the surface. That's not the only remarkable thing about these complex, fascinating fish.

Coping with potentially hostile environments is a major key to success in the fish world, and the labyrinth fish (popularly known as anabantids) have evolved physical characteristics and breeding techniques that place them high in the survival stakes. The suborder to which they belong, Anabantoidei, includes five families, but only the family Belontidae (Gouramis, Bettas, Combtails and Paradisefish) is well represented in the fishkeeping hobby. Most of these fish are small, vivid, well-behaved and relatively easy to keep and breed. Anabantids come from Africa and Asia; most hobby fish are imported from the Far East.

The labyrinth is a respiratory organ located above each gill chamber, enabling anabantids to breathe atmospheric oxygen. Air is gulped in at the surface and forced into the labyrinth, which is furnished with a rich capillary blood supply. Oxygen diffuses into the bloodstream and carbon dioxide is unloaded before the spent air is expelled through the

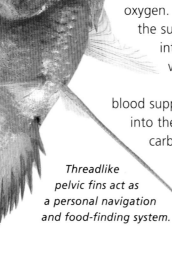

Threadlike pelvic fins act as a personal navigation and food-finding system.

Anabantids gulp surface air to fill their labyrinth organs. These spongy outgrowths of the gill chamber act like lungs to boost oxygen levels in the blood. By contrast, these fish blow bubbles during nest building.

gills. As a result, these fish can inhabit roadside ditches and rice paddies, where the oxygen-deficient and often polluted waters would prove fatal to most others. Some Gouramis even have greatly extended, threadlike pelvic fins to "feel" their way through the murk and find food (small aquatic insects and crustaceans). The two modes of breathing complement one another, so that if an anabantid is unable to visit the surface at regular intervals, it will die. The other characteristic survival technique relates to breeding. Bubble-nesting, with male parental care, gives fry a good chance of avoiding predation in the critical first days of life, and therefore broods are typically smaller than those of egg-scatterers. Bubble-nests are constructed by the males and vary in size and complexity according to species. Some incorporate plant matter; others rely purely on a film of saliva around each bubble to hold them together. On completion of the nest, the pair entwine below it in a spawning embrace and eggs are released and fertilized.

A male Betta spreads his fins to challenge a rival head-on.

These are gathered by the male, who spits them into their blanket of bubbles and guards them until they hatch. With Gouramis of the genus *Trichogaster* the eggs float, but with Bettas *(Betta splendens)* they sink, and the male is kept busy returning them to the nest. Newly hatched fry take an initial gulp of surface air to fill their swim bladders. This air is warm and humid in the wild, so for successful aquarium spawnings, keep the tank covered.

BETTA

Betta splendens

Mature size: 2.75 inches (7 cm)

Characteristics: Devoted parent but merciless to other males, the male Betta (also sold as Siamese Fighting Fish) lives its short life on the edge. Flowing fins and flared gill covers (below) are displayed as threat gestures that quickly develop into ferocious attacks on each other's fins. A sole male will

live peacefully in a community aquarium, but a breeding pair need a well-planted tank to themselves. Even then, the testosterone-fueled male can turn on his plainer mate. As with other bubble-nesters, the male guards the eggs floating at the surface. Selective breeding has produced ever more extravagant finnage in shimmering combinations of red, blue, green and gold.

BETTA

Betta splendens

Mature size: 2.75 inches
(7 cm)

Characteristics: In repose, this
male steel-blue Betta shows off his
beautiful trailing finnage, which is far
more elaborate than would be found in
the wild fish. How long will he be around
to flaunt his finery? In an aquarium, even
without the presence of a female and the
pressures of fatherhood to keep him
running on adrenalin, two years is a good
age for a Betta. His life is high-octane but
short, so treat him well. Do not mix him
with fin-nippers or taunt him with rival
males in nearby tanks, otherwise he will
conduct his aggression dramas through
the glass and burn out like the fiery
living candle he is.

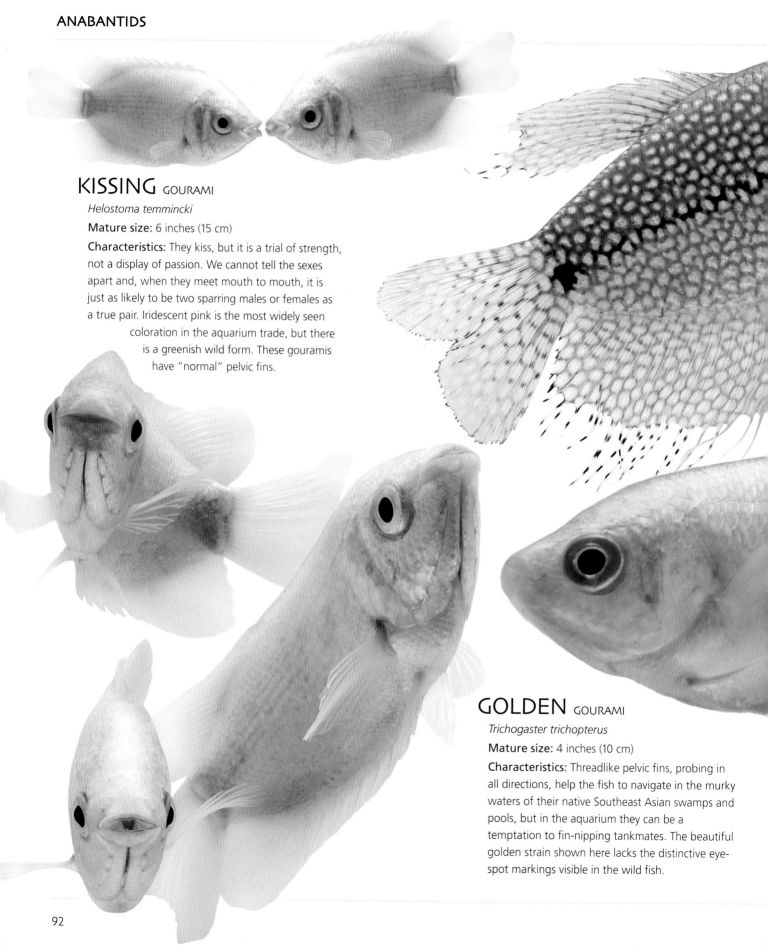

KISSING GOURAMI

Helostoma temmincki

Mature size: 6 inches (15 cm)

Characteristics: They kiss, but it is a trial of strength, not a display of passion. We cannot tell the sexes apart and, when they meet mouth to mouth, it is just as likely to be two sparring males or females as a true pair. Iridescent pink is the most widely seen coloration in the aquarium trade, but there is a greenish wild form. These gouramis have "normal" pelvic fins.

GOLDEN GOURAMI

Trichogaster trichopterus

Mature size: 4 inches (10 cm)

Characteristics: Threadlike pelvic fins, probing in all directions, help the fish to navigate in the murky waters of their native Southeast Asian swamps and pools, but in the aquarium they can be a temptation to fin-nipping tankmates. The beautiful golden strain shown here lacks the distinctive eye-spot markings visible in the wild fish.

PEARL GOURAMI

Trichogaster leeri

Mature size: 4 inches (10 cm)

Characteristics: The male, his rosy body shot with pearly iridescence, is in full breeding dress. He will build a floating bubble-nest before meeting in a spawning embrace with his partner, whose coloration is less intense and whose finnage is not quite as flamboyant. Then he will be a model parent to the eggs and tiny fry.

RED DWARF
GOURAMI

Colisa lalia

Mature size: 2 inches (5 cm)

Characteristics: Two species on this page are aquarium variants of the Dwarf Gourami, in which the males show vertical flank barring and blue-and-red spangling on the body and fins, while females are plainer and more silvery. In the Red Dwarf only the more pointed dorsal of the male is a sure way of identifying the sexes. The humped shoulder profile is typical of Dwarf Gouramis. They are peaceable, if somewhat shy.

♂

♀

BANDED GOURAMI

Colisa fasciata

Mature size: 4 inches (10 cm)

Characteristics: A previous alternative name for this fish was Giant Gourami, but all things are relative; today the term is applied only to a true leviathan, *Osphronemus gorami*, which can reach 27 inches (70cm)! The Banded Gourami can lay up to 1,000 eggs — hard work for a dutiful male. Eye-catching features are the red-edged, electric blue anal fin and the prominent red eye.

BLUE DWARF
GOURAMI

Colisa lalia

Mature size: 2 inches (5 cm)

Characteristics: In this aquarium strain, blue is the dominant coloration, although you can still see the flank bars showing through. The upturned mouth is well adapted to feeding on insects at the water surface and (in the male) for taking air into the labyrinth organ and blowing a bubble-nest. All Gouramis benefit from mature water, so wait a while before introducing them to your tank.

♂

♀

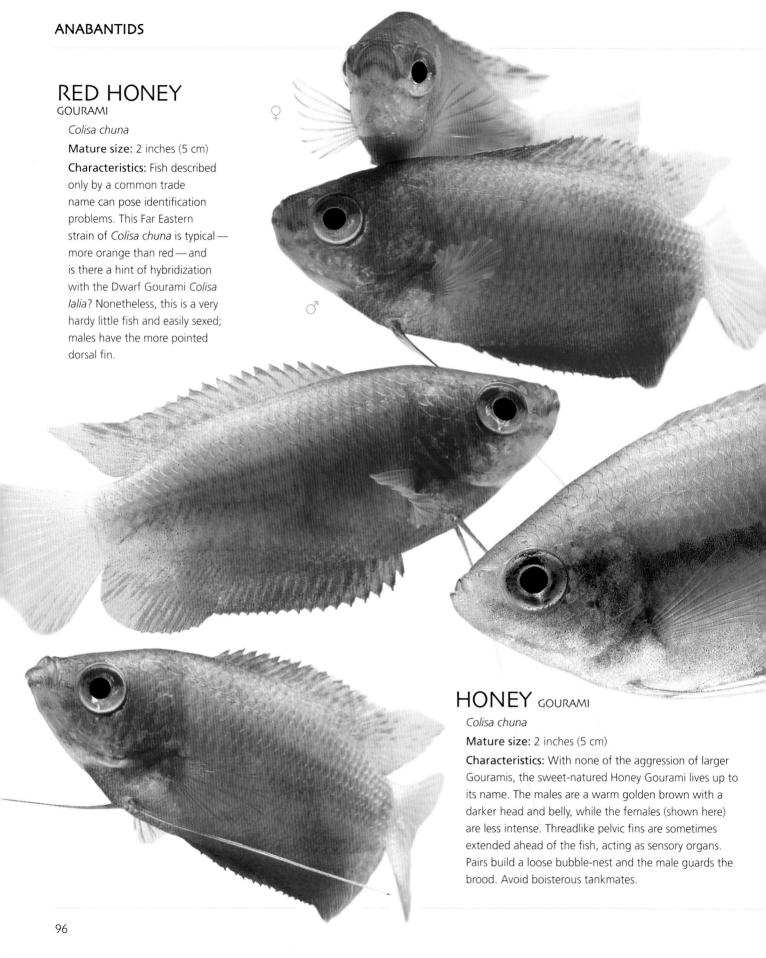

RED HONEY
GOURAMI

Colisa chuna

Mature size: 2 inches (5 cm)

Characteristics: Fish described only by a common trade name can pose identification problems. This Far Eastern strain of *Colisa chuna* is typical — more orange than red — and is there a hint of hybridization with the Dwarf Gourami *Colisa lalia*? Nonetheless, this is a very hardy little fish and easily sexed; males have the more pointed dorsal fin.

♀

♂

HONEY GOURAMI

Colisa chuna

Mature size: 2 inches (5 cm)

Characteristics: With none of the aggression of larger Gouramis, the sweet-natured Honey Gourami lives up to its name. The males are a warm golden brown with a darker head and belly, while the females (shown here) are less intense. Threadlike pelvic fins are sometimes extended ahead of the fish, acting as sensory organs. Pairs build a loose bubble-nest and the male guards the brood. Avoid boisterous tankmates.

OPALINE GOURAMI

Trichogaster trichopterus

Mature size: 4 inches (10 cm)

Characteristics: The wild fish, also known as the Three-spot Gourami, is mottled blue, with two black spots along the body (the eye being the third spot). Aquarium strains such as this result from crosses with *Trichogaster trichopterus trichopterus*, which grows slightly larger. Females become noticeably plump when ready to spawn. These Gouramis are quite peaceful, although the occasional rogue male can harass smaller tankmates.

DANIOS, & RASBORAS

The Scissortail Rasbora scores highly in the small-fish popularity stakes.

Sooner or later, most aquarists like to increase their stocks by breeding their own fish. Enter on cue the Zebra Danio *(Brachydanio rerio)*, the perfect egg-layer for beginners. This little fish from India is easy to sex (females are more vivid, males slimmer), comes readily into condition and will spawn virtually to order. I remember the first time I saw a pin-striped pair placed into a small breeding tank, bare except for a layer of glass marbles on the bottom. A measure of cold water dropped the temperature a couple of degrees and almost instantly the fish began a whirling courtship dance, too fast for the eye to follow. Tiny, clear fertilized eggs fell into the crevices between the marbles, the

parents were removed, and a couple of days later the first fry were seen clinging like tiny glass splinters to the walls of the tank.

The only other egg-layer remotely as obliging is the White Cloud Mountain Minnow *(Tanichthys albonubes)*, a small, cool-water-loving cyprinid originally from China.

"Danio" describes fish of two genera, *Brachydanio* and *Danio*, the difference being that *Brachydanio* has an incomplete lateral line (the sensory organ along the flanks). From a fishkeeping viewpoint, they can all be treated the same. All danios are from Southeast Asia and are

Lively Gold Danios lack the blue pigment of their Zebra forebears.

MINNOWS

members of the carp family, Cyprinidae. They are streamlined, fast-swimming shoaling fish that occupy the often underpopulated middle layers of the aquarium. They pose no problems to tankmates and are agile enough to evade casual harassment.

The Zebra Danio is among the least expensive of tropical fish, extensively farmed, with long-finned variants common. The Pearl Danio *(Brachydanio albolineatus)* is a slightly larger fish with an iridescent sheen. The Leopard Danio *(Brachydanio frankei)* is rather a mystery — some consider it a distinct species, others a sport of the Zebra Danio. The gold body is dappled with blue, a pattern that runs into the anal and caudal fins.

Two species of true Danio are suited to larger, planted aquariums. The Giant *(Danio aequipinnatus)* and Bengal Danio *(Danio devario)* are both fish of running, rather than stagnant, water and require good aeration.

Rasboras are less easy to breed than danios, but otherwise are very similar in disposition. They are found from East Africa across eastern Asia to the Philippines and Indonesia. Harlequin, Clown and Scissortail Rasboras are justifiably popular aquarium fish. To encourage spawning, the water needs to be soft and acidic. Adults adapt to tap-water close to neutral.

A plump female Harlequin (Rasbora heteromorpha). *In males, the vertical edge of the dark marking is sloping.*

Spot the link — just join the dots and a Leopard Danio (top) becomes a Zebra. The two are clearly closely linked.

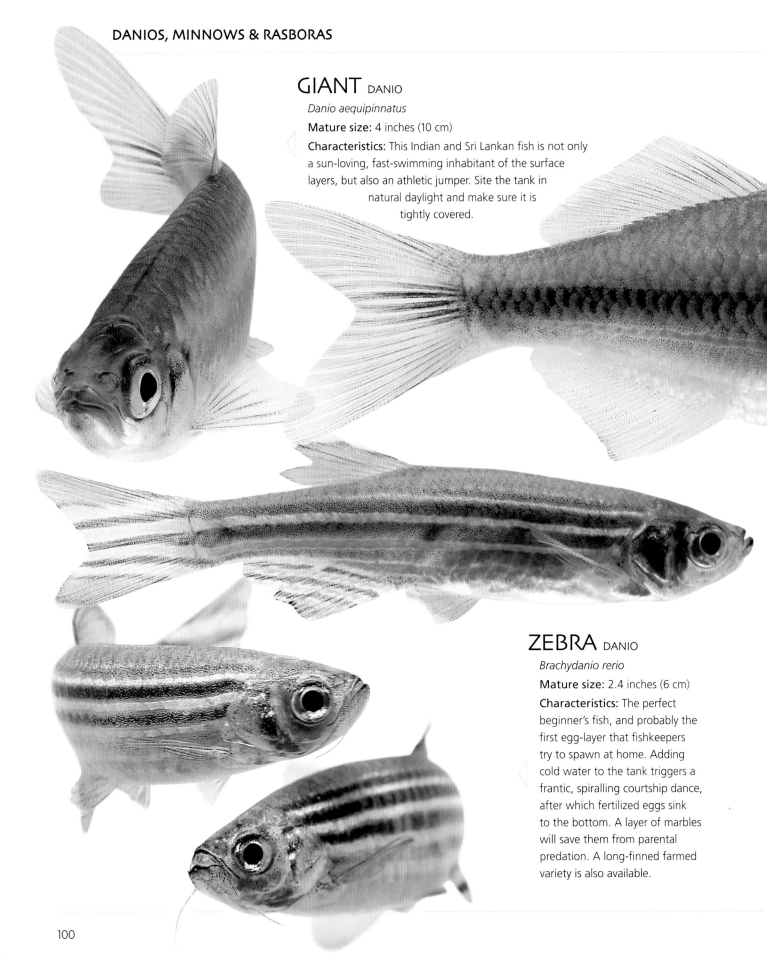

GIANT DANIO

Danio aequipinnatus

Mature size: 4 inches (10 cm)

Characteristics: This Indian and Sri Lankan fish is not only a sun-loving, fast-swimming inhabitant of the surface layers, but also an athletic jumper. Site the tank in natural daylight and make sure it is tightly covered.

ZEBRA DANIO

Brachydanio rerio

Mature size: 2.4 inches (6 cm)

Characteristics: The perfect beginner's fish, and probably the first egg-layer that fishkeepers try to spawn at home. Adding cold water to the tank triggers a frantic, spiralling courtship dance, after which fertilized eggs sink to the bottom. A layer of marbles will save them from parental predation. A long-finned farmed variety is also available.

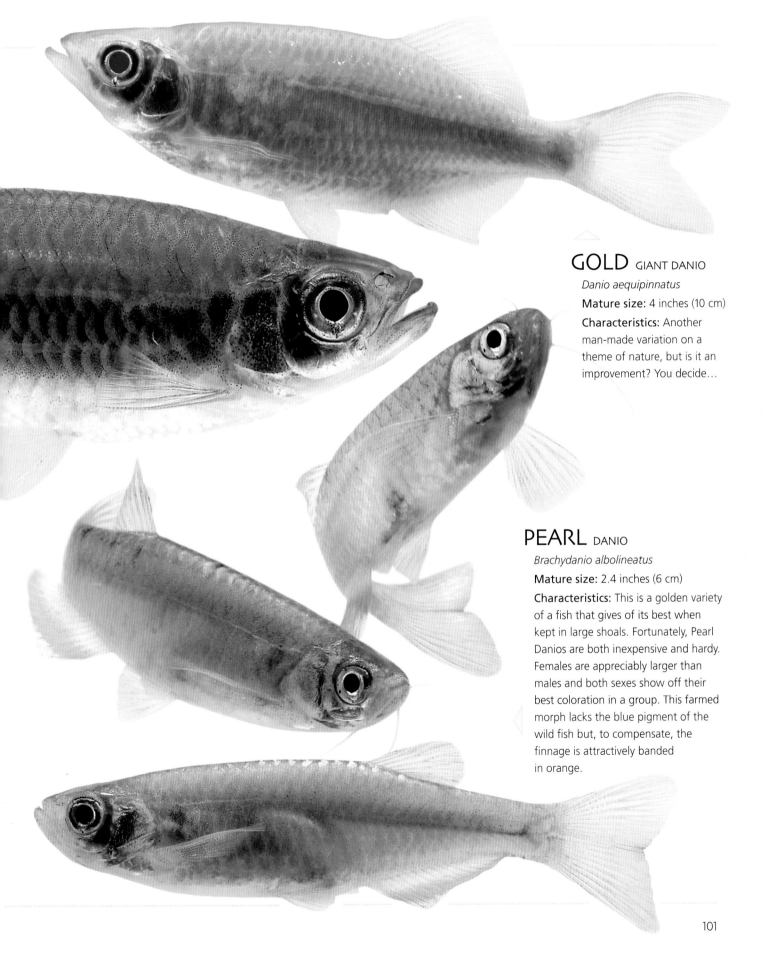

GOLD GIANT DANIO

Danio aequipinnatus

Mature size: 4 inches (10 cm)

Characteristics: Another man-made variation on a theme of nature, but is it an improvement? You decide…

PEARL DANIO

Brachydanio albolineatus

Mature size: 2.4 inches (6 cm)

Characteristics: This is a golden variety of a fish that gives of its best when kept in large shoals. Fortunately, Pearl Danios are both inexpensive and hardy. Females are appreciably larger than males and both sexes show off their best coloration in a group. This farmed morph lacks the blue pigment of the wild fish but, to compensate, the finnage is attractively banded in orange.

WHITE CLOUD MOUNTAIN
MINNOW

Tanichthys albonubes

Mature size: 1.6 inches (4 cm)

Characteristics: These southern Chinese natives appreciate a cool aquarium—no more than 72°F (22°C). Although a long-finned variety is available, nothing matches a shoal of the originals for subdued chic.

HARLEQUIN RASBORA

Rasbora heteromorpha

Mature size: 1.8 inches (4.5 cm)

Characteristics: At their best in large shoals with equally peaceable fish, these fish spawn on the undersides of broad-leaved aquarium plants. Females are deeper-bodied, with a straight front edge to the black marking.

SCISSORTAIL RASBORA

Rasbora trilineatus

Mature size: 6 inches (15 cm)

Characteristics: One of the larger rasboras, also known as "Three-lined," these fish hail from Malaysia, Sumatra and Borneo. Are those prominent tail markings a means of escaping predators with minimal collateral damage? That is where the eye of a piscivore would be drawn.

Shoaling or solitary, dwarf or giant, catfish truly offer something for everyone, even though their bizarre appearance and way of life often challenges belief.

Catfish are incredibly diverse and wide-ranging. It is hard to believe that an inoffensive 2-inch (5 cm) algae-eater from Brazil belongs to the same group as the glassy-eyed, fearsome 5-foot (1.5m) predator *Hemibagrus wyckii* from Southeast Asia, whose velvet-black skin matches its dark disposition. And yet, when you realize that

catfish have successfully colonized all the tropical and temperate regions, it is clear that they have had to become masters of habitat exploitation.

What is a catfish? There are 34 catfish families, including marine and brackish-water species, and common to them all is a lack of scales. Either their skins are naked or they are furnished with bony plates. Usually, but not always, it is the

The omega eye in the Royal Pleco (Panaque nigrolineatus) *shades the sensitive retina from bright sunlight.*

smaller prey catfish that are protected in this way. Catfish are so named because they have "whiskers," although in some species these are

inconspicuous. These nerve-rich paired sensory organs — known as barbels — are used to locate food and, in murky waters, for navigation. In some bagrids and pimelodids, they can be extravagantly long, sweeping almost the length of the body. In *Corydoras* catfish, they are relatively short, while the

Corydoras (here, C. gossei) *are mistakenly imagined to be the scavengers among catfish.*

H

Transparently beautiful — the African Glass Catfish (Kryptopterus bicirrhis).

bristlenose cats have only one pair of true barbels at the corners of their mouths, supplemented by fleshy outgrowths that are more prominent in the males.

Quirky lifestyles endear cats to fishkeepers. The African *Synodontis nigriventris* routinely swims upside down — hence its common name of Upside-down Catfish — while its relative *S. multipunctatus*, known as the Cuckoo

Catfish, sneaks its eggs into the broods of mouthbreeding cichlids, which pick them up and rear them as their own. The Walking Catfish, *Clarias batrachus,* is undismayed when its Malaysian home waters dry up; "breathing" air with a primitive lung, it sets off across dry land, moving fast on its pectoral fin spines.

Catfish can "talk," too: lift the South American Talking Catfish, *Amblydoras hancockii,* out of the water and it will grunt at you.

In the brood-care stakes, catfish run the gamut of egg-scatterers, mouthbreeders and cave-spawners; in all cases,

the male guards the young. There are bubble-nest-builders too, such as *Callichthys callichthys*. With some Asian catfish, brood care is shared. In these nuclear families, the male guards the eggs and both parents then look after the newly hatched youngsters until they can fend for themselves. With *Corydoras* catfish, the most popular of all aquarium catfish, the female picks up fertilized eggs in a cup formed by her ventral fins and then sticks them to a flat surface, usually a plant or the tank glass.

The splendid barbels of Synodontis eupterus.

ALBINO BRONZE CORY

Corydoras aeneus

Mature size: 2.75 inches (7 cm)

Characteristics: Albino corys were originally spontaneous mutations lacking black pigment, but are now bred to order.

BRONZE CORY

Corydoras aeneus

Mature size: 2.75 inches (7 cm)

Characteristics: Fish do not have eyelids, but the rapid rotation of the eyeballs in their sockets, a typical *Corydoras* trait, gives the rather endearing illusion that they are winking, adding to the general cuteness of these inoffensive little catfish.

PERUVIAN
GREEN STRIPE CORY

Corydoras cf. *aeneus*

Mature size: 2.75 inches (7 cm)

Characteristics: This "strain" of the Bronze Cory is one of the gems among what is normally considered a rather drab fish. The enameled scales on the flanks reflect the light; the long, undamaged barbels are a sign of fish in vigorous good health.

SUNSET CORY

Corydoras sp.

Mature size: 2.75 inches (7 cm)

Characteristics: Feeding to improve
natural coloration may have an
influence on these startlingly bright
corys, and this is a perfectly
acceptable practice. Not so is the
injection of dye into mass-farmed
tropical fish, though it still goes on,
particularly with Indian Glassfish.
If you suspect that your aquatic
store knowingly stocks dyed fish,
take your business elsewhere.
Nature's diversity should not
be abused in this way.

PANDA CATFISH

Corydoras panda

Mature size: 2 inches (5 cm)

Characteristics: It is easy to see how this appealing little Peruvian catfish got its name—the black mask, running from the top of the head down through the eyes, lends it a very pandalike appearance, especially when viewed head-on. To maintain their barbels in good condition, all *Corydoras* should be kept in an aquarium with a sandy, rather than a gravel, substrate; otherwise these nerve-rich sensory organs will be worn down.

♂   ♀

DUPLICATE CORY

Corydoras duplicareus

Mature size: 2.4 inches (6 cm)

Characteristics: Members of the masked *Corydoras* group look very similar, and this one was named for its resemblance to *C. adolfoi*, also from Brazil, which was described first. If anything, the mask is even more pronounced than that of *C. panda*, and the black markings extend all along the dorsal surface, setting off the white flanks with a marked greenish iridescence.

STERBA'S CORY

Corydoras sterbai

Mature size: 2.4 inches (6 cm)

Characteristics: Vertical bars on the caudal fin, contrasting horizontal lines of black spots on the flanks and a neatly spotted head — look closely and be amazed at the complexity of the pattern of this Brazilian cory. As in all *Corydoras*, ripe females become quite rotund compared to males.

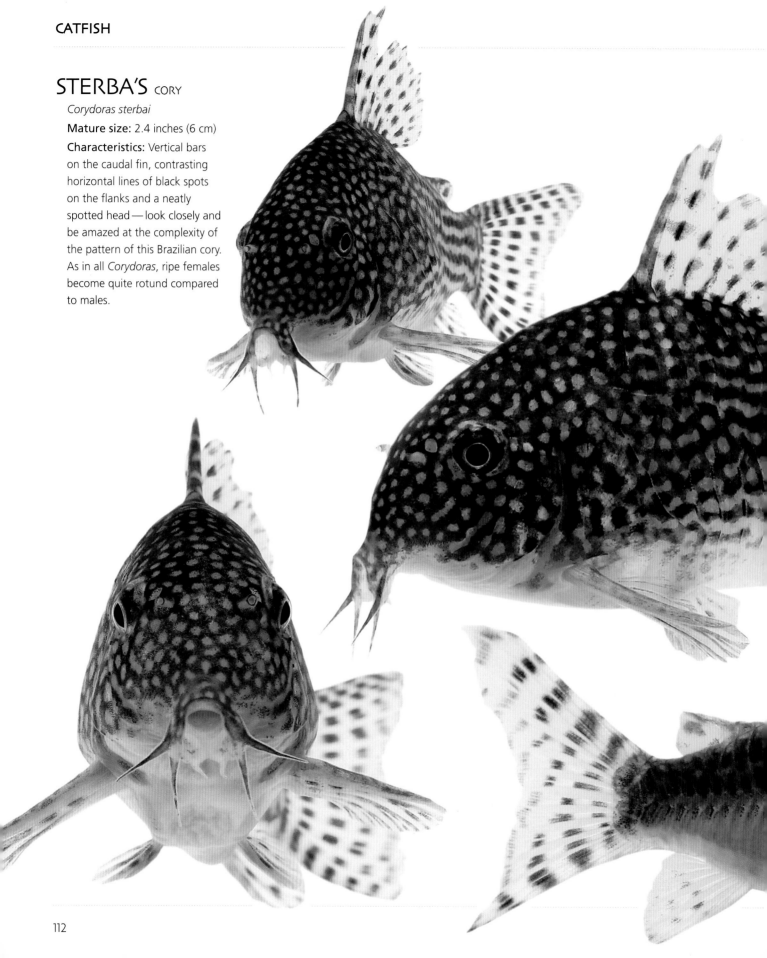

With space and good food, these young could reach breeding size within six to eight months.

GOSSE'S CORY

Corydoras gossei

Mature size: 2 inches (5 cm)

Characteristics: *Corydoras*, the vacuum cleaners of the catfish world, were once wrongly thought to scavenge leftovers from their tankmates. Now they are appreciated in their own right. This chunky little character, with its blunt head and surprised expression, could live for decades if given a suitably varied diet of live and flake food.

EMERALD CATFISH

Brochis splendens

Mature size: 2.75 inches (7 cm)

Characteristics: The longer dorsal fin is an instant giveaway, showing that these callichthyids are not true *Corydoras*, although they are very closely related. *Brochis* are more robustly built, deeper in the body, with a more massive head and a longer snout, but in temperament there are no real differences between the genera. They lay large, adhesive eggs on flat vertical or horizontal surfaces.

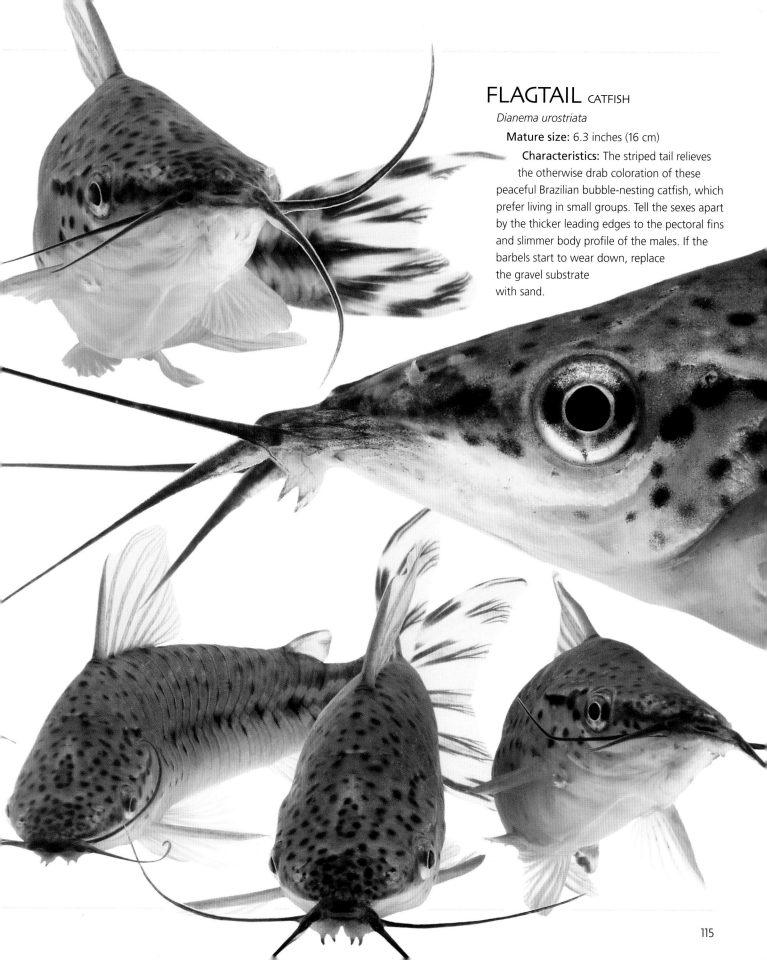

FLAGTAIL CATFISH

Dianema urostriata

Mature size: 6.3 inches (16 cm)

Characteristics: The striped tail relieves the otherwise drab coloration of these peaceful Brazilian bubble-nesting catfish, which prefer living in small groups. Tell the sexes apart by the thicker leading edges to the pectoral fins and slimmer body profile of the males. If the barbels start to wear down, replace the gravel substrate with sand.

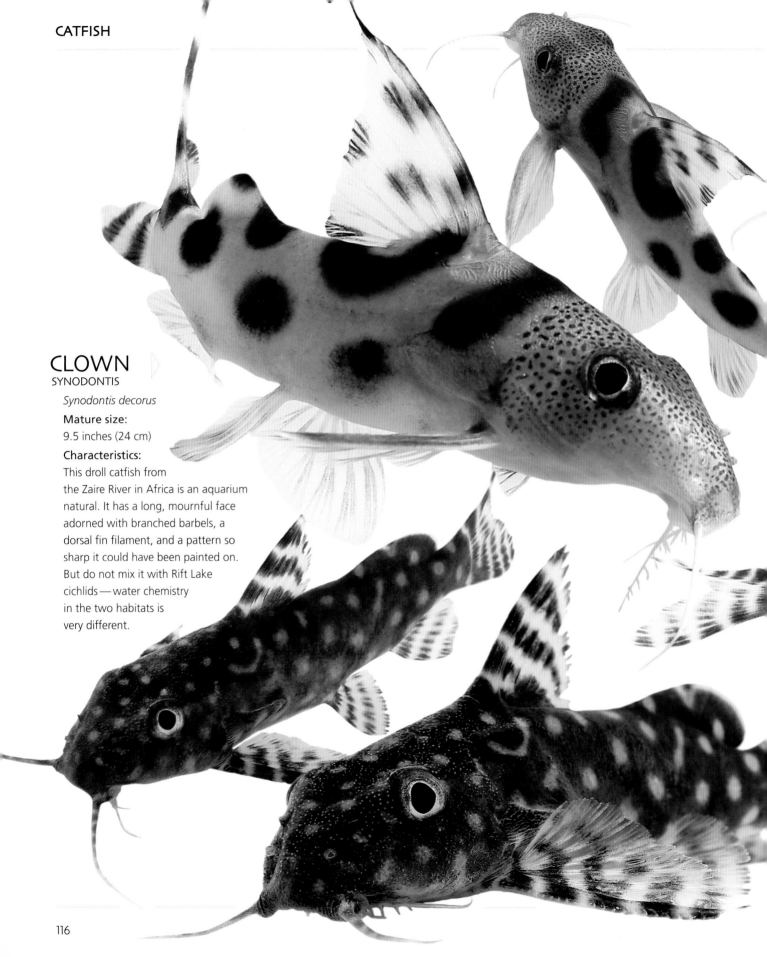

CLOWN
SYNODONTIS
Synodontis decorus

Mature size:
9.5 inches (24 cm)

Characteristics:
This droll catfish from
the Zaire River in Africa is an aquarium
natural. It has a long, mournful face
adorned with branched barbels, a
dorsal fin filament, and a pattern so
sharp it could have been painted on.
But do not mix it with Rift Lake
cichlids—water chemistry
in the two habitats is
very different.

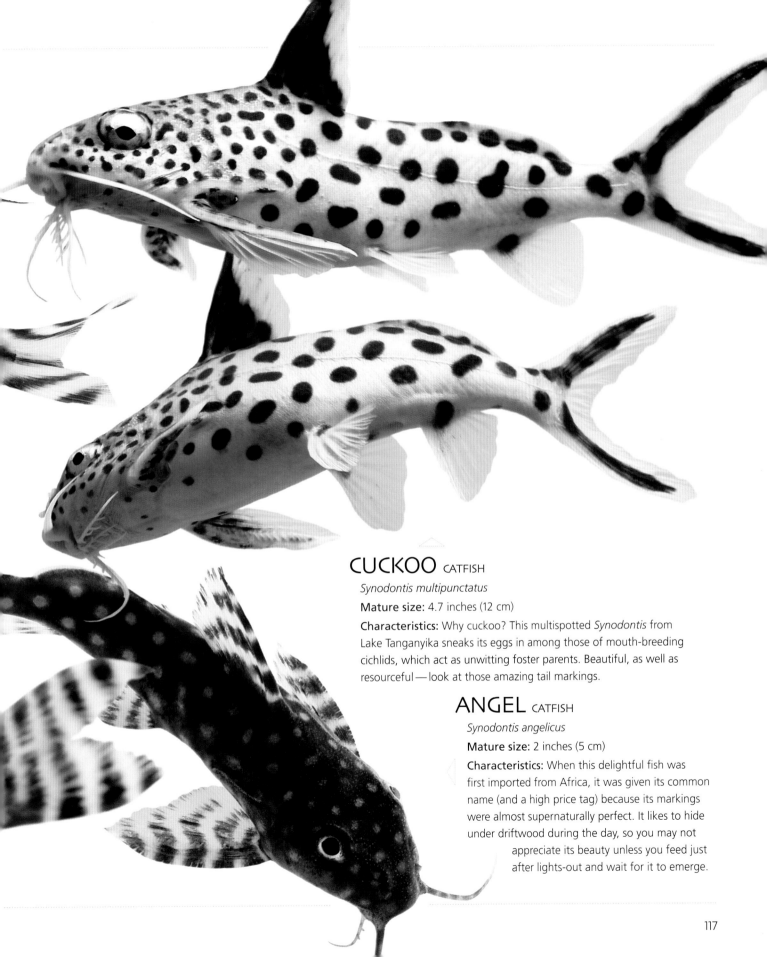

CUCKOO CATFISH

Synodontis multipunctatus

Mature size: 4.7 inches (12 cm)

Characteristics: Why cuckoo? This multispotted *Synodontis* from Lake Tanganyika sneaks its eggs in among those of mouth-breeding cichlids, which act as unwitting foster parents. Beautiful, as well as resourceful—look at those amazing tail markings.

ANGEL CATFISH

Synodontis angelicus

Mature size: 2 inches (5 cm)

Characteristics: When this delightful fish was first imported from Africa, it was given its common name (and a high price tag) because its markings were almost supernaturally perfect. It likes to hide under driftwood during the day, so you may not appreciate its beauty unless you feed just after lights-out and wait for it to emerge.

The two fish shown here are the same species, but clearly they show different coloration — probably as a result of how they have been kept. The paler fish has responded to bright light in its aquarium by contracting the pigment cells (melanophores) in the skin. The darker fish may have been kept in subdued light or with a dark substrate, making the pigment cells expand. Mood changes, stress or readiness to spawn are other reasons for these changes.

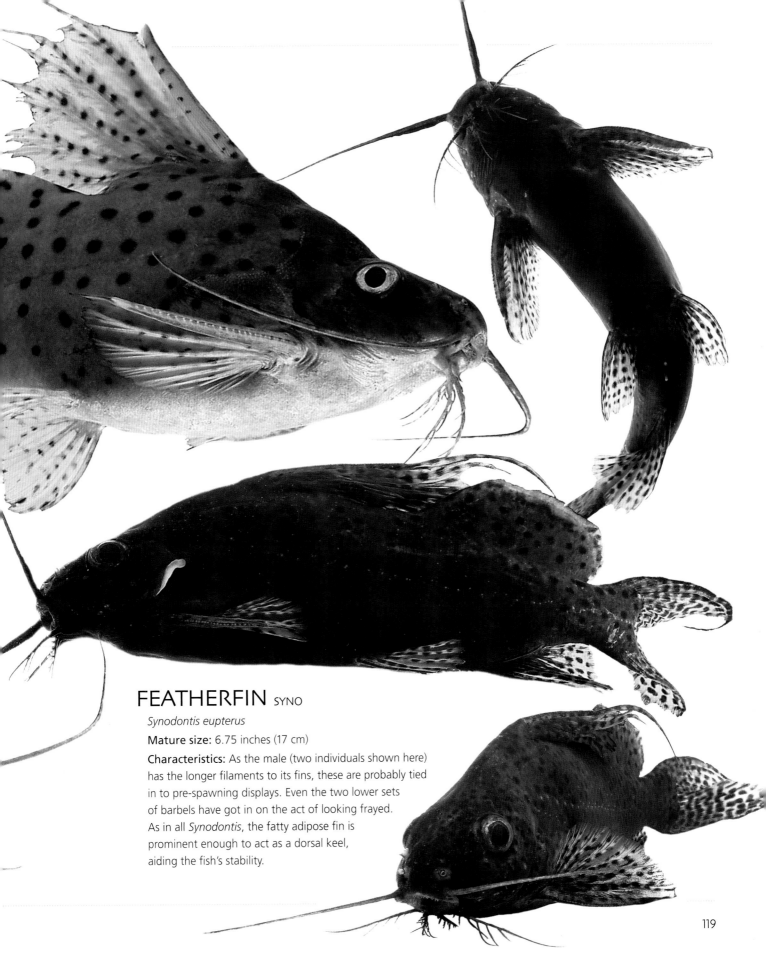

FEATHERFIN SYNO

Synodontis eupterus

Mature size: 6.75 inches (17 cm)

Characteristics: As the male (two individuals shown here) has the longer filaments to its fins, these are probably tied in to pre-spawning displays. Even the two lower sets of barbels have got in on the act of looking frayed. As in all *Synodontis*, the fatty adipose fin is prominent enough to act as a dorsal keel, aiding the fish's stability.

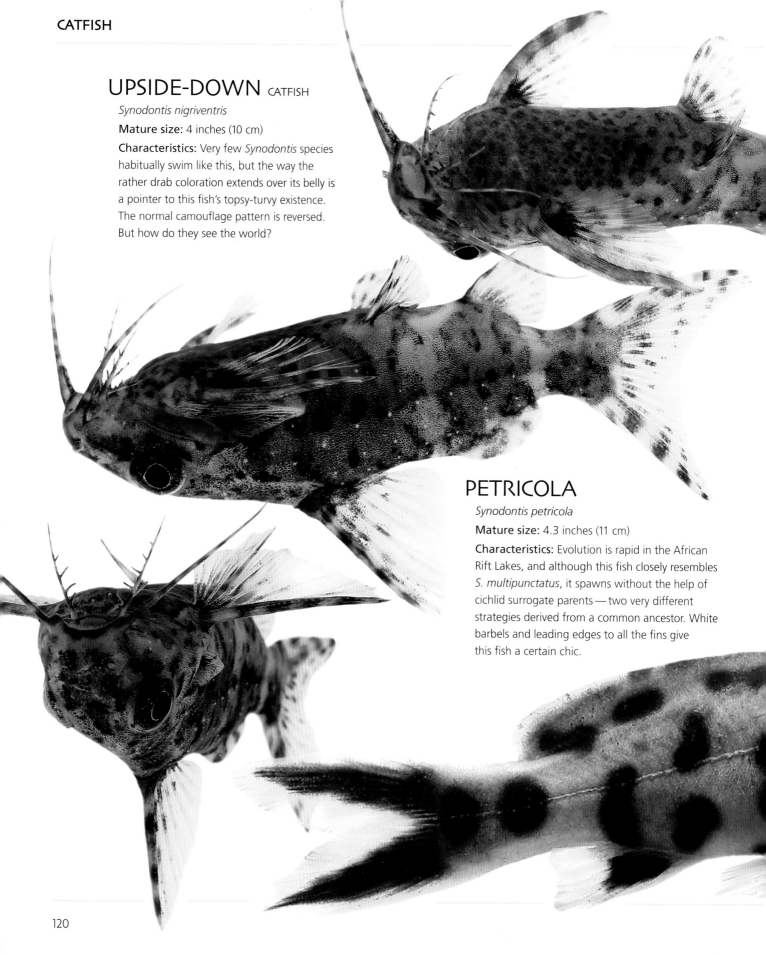

UPSIDE-DOWN CATFISH

Synodontis nigriventris

Mature size: 4 inches (10 cm)

Characteristics: Very few *Synodontis* species habitually swim like this, but the way the rather drab coloration extends over its belly is a pointer to this fish's topsy-turvy existence. The normal camouflage pattern is reversed. But how do they see the world?

PETRICOLA

Synodontis petricola

Mature size: 4.3 inches (11 cm)

Characteristics: Evolution is rapid in the African Rift Lakes, and although this fish closely resembles *S. multipunctatus*, it spawns without the help of cichlid surrogate parents — two very different strategies derived from a common ancestor. White barbels and leading edges to all the fins give this fish a certain chic.

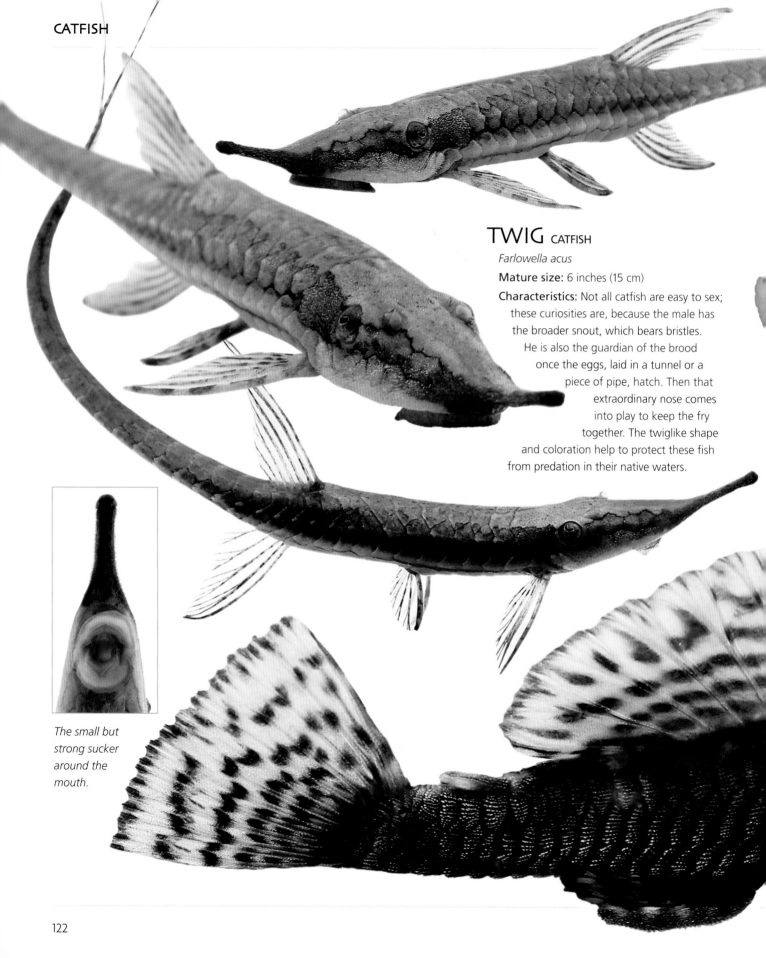

TWIG CATFISH

Farlowella acus

Mature size: 6 inches (15 cm)

Characteristics: Not all catfish are easy to sex; these curiosities are, because the male has the broader snout, which bears bristles. He is also the guardian of the brood once the eggs, laid in a tunnel or a piece of pipe, hatch. Then that extraordinary nose comes into play to keep the fry together. The twiglike shape and coloration help to protect these fish from predation in their native waters.

The small but strong sucker around the mouth.

These views show the sucker mouth on the leucistic (left) and normal forms.

BRISTLENOSE
CATFISH

Ancistrus temminckii

Mature size: 4.7 inches (12 cm)

Characteristics: To a female *Ancistrus*, those fleshy nose bristles signal "macho mate" — and this fearsome-looking mini combat tank proves a great dad too, brooding a clutch of orange eggs under a rock or in a cave. These are the easiest sucker-mouths to spawn in an aquarium. The top fish is a leucistic (semi-albino) form.

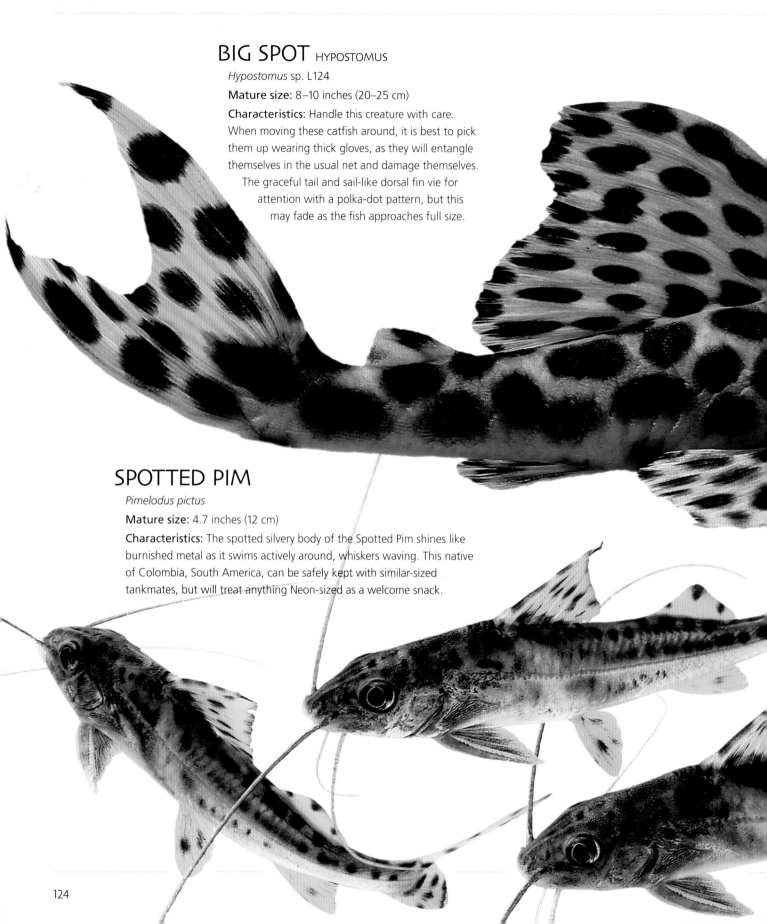

BIG SPOT HYPOSTOMUS

Hypostomus sp. L124

Mature size: 8–10 inches (20–25 cm)

Characteristics: Handle this creature with care. When moving these catfish around, it is best to pick them up wearing thick gloves, as they will entangle themselves in the usual net and damage themselves. The graceful tail and sail-like dorsal fin vie for attention with a polka-dot pattern, but this may fade as the fish approaches full size.

SPOTTED PIM

Pimelodus pictus

Mature size: 4.7 inches (12 cm)

Characteristics: The spotted silvery body of the Spotted Pim shines like burnished metal as it swims actively around, whiskers waving. This native of Colombia, South America, can be safely kept with similar-sized tankmates, but will treat anything Neon-sized as a welcome snack.

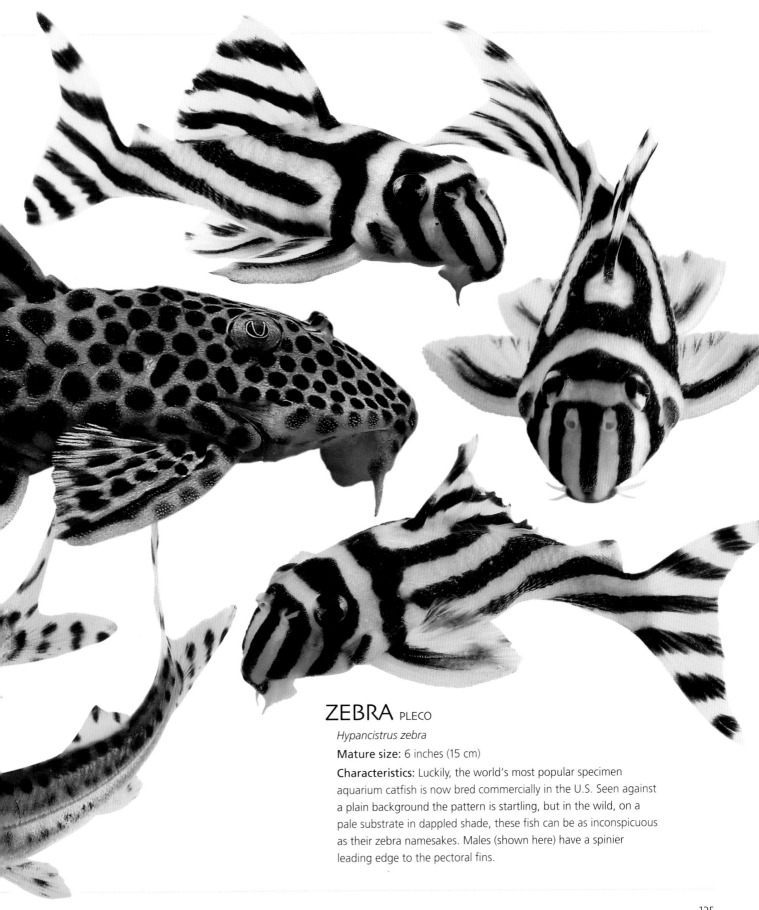

ZEBRA PLECO

Hypancistrus zebra

Mature size: 6 inches (15 cm)

Characteristics: Luckily, the world's most popular specimen aquarium catfish is now bred commercially in the U.S. Seen against a plain background the pattern is startling, but in the wild, on a pale substrate in dappled shade, these fish can be as inconspicuous as their zebra namesakes. Males (shown here) have a spinier leading edge to the pectoral fins.

GOLDY PLECO

Scobinancistrus aureatus

Mature size: 12 inches (30 cm)

Characteristics: So many similar-looking South American sucker-mouths are being imported these days that making a positive identification can be very difficult. To assist, many of them carry L, or location, numbers that pinpoint where they were collected, and this one is L14, from Brazil's Rio Xingu. It is hard not to smile as you look at this fish, which seems to be bathed in golden sunlight. This example, with the dappled body pattern extending well into the fins, shows just how protrusible the sucker-mouth is. The single pair of barbels on the outer corners of the lips are quite elongated, as you would expect of an animal from a fast-flowing river, where sensory information is carried rapidly down on the current and needs to be acted upon quickly if the fish is to obtain a meal.

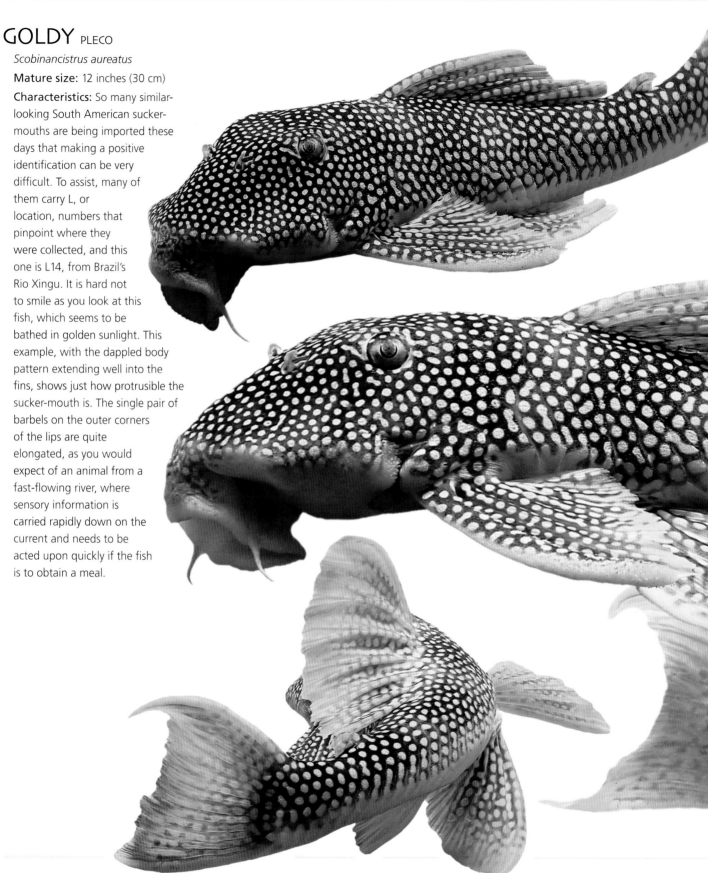

REDTAIL CATFISH
Phractocephalus hemioliopterus

Mature size: 72 inches (180 cm)

Characteristics: One to admire in public aquariums, rather than try to keep at home — a cute baby is one thing, a 6-foot predator quite another. But there is no denying the "wow" factor of this creature, with its go-faster flank stripe, reddish orange propulsion unit and three wonderfully mobile pairs of barbels.

CACTUS PLECO
Pseudacanthicus sp. L25

Mature size: 12 inches (30 cm)

Characteristics: Once, all unidentified sucker-mouth catfish were labelled plecos, but accurate identification is the key to adult size. The thorny and impregnable body protection of this fish explains its common name.

KING GALAXY PLECO

Pseudacanthicus sp. L282

Mature size: 8 inches (20 cm)

Characteristics: A recent import from Brazil, identified only by its L number.
What a beauty it is, from all angles!

CRYSTAL-EYE CATFISH

Hemibagrus wyckii

Mature size: 32 inches (80 cm)

Characteristics: A Southeast Asian villain, jet-black, predatory and built for speed. Keep it as a pet in a tank on its own, but mind your fingers at feeding time!

ROYAL PLECO

Panaque nigrolineatus

Mature size: 10 inches (25 cm)

Characteristics: Startling zebralike facial stripes and curious omega-shaped pupils make this a show-stopping aquarium star. Rasping away at pieces of driftwood aquarium decor aids its digestion, but do not expect it to clear the tank of algae. It will much prefer sinking catfish pellets or a piece of cucumber.

Placid, adaptable and diverse in body shape and coloration, exotic rainbowfish bring gasps of admiration from all who encounter them for the first time.

Adapting tropical fish to the local tap water can be difficult: although many farmed community species (including South American tetras) will live quite happily in moderately hard, alkaline water, they need soft, acidic water if they are to breed successfully.

This is not the case with Rainbowfish, the chemistry of whose native Australasian waters closely matches what comes out of the tap in areas where the supply is drawn from chalk or limestone rocks. That, teamed with their beauty and peaceful nature, means Rainbowfish are enjoying a surge in popularity.

There are eight families in the order Atheriformes, two of which are exclusively marine fish; from an aquarium point of view only three others are of interest. Most species we see for sale belong to the family Melanotaenidae, from northern and eastern Australia and New Guinea. The two exceptions

Slim, subtle and streamlined sum up the elegant Threadfin Rainbowfish.

The Salmon Red Rainbow is deep-bodied, with simple, vivid coloration.

are the Madagascar Rainbowfish *(Bedotia geayi)* and the Celebes Rainbow *(Marosatherina ladigesi)* from Sulawesi.

A double dorsal fin is a characteristic of Rainbowfish.

The longer, hindmost fin closely matches the anal fin in shape and coloration, giving the fish a pleasing symmetry. This is carried to extremes in the Threadfin Rainbow *(Iriatherina werneri),*

W FISH

males of which have long fin extensions used to display to potential mates.

Very few wild-caught Rainbowfish are offered for sale and, unfortunately, the stunning coloration becomes muted after a few generations of captive breeding. This, and the existence of local subspecies, can make positive identification difficult. Nonetheless, the male Salmon Red Rainbow (*Glossolepis incisus*) remains a particularly striking fish, carrying the deep crimson body shade into all its finnage. It makes a superb occupant of display aquariums, where it occupies the middle layers of the water.

Rainbowfish look best in large, well-planted tanks with a dark sandy substrate and some spotlighting supplementing the overhead tubes. Their coloration

Male Celebes Rainbowfish with yellow fins and noses.

appears to change and shimmer as they swim from shaded into brighter areas, hence their common name.

Although *Melanotaenia* females tend to be less vivid, keeping both sexes together encourages the males to show off their best. Like the tetras, they are happiest living in shoals in the aquarium.

They will eat all types of commercially prepared foods, although their diet in the wild is made up of insects, small aquatic invertebrates and sometimes some algae.

Rainbowfish spawnings tend to be protracted, a few eggs being laid each day over a long period in plant thickets or artificial spawning mops. The fry need very fine first food, such as newly hatched brine shrimp (*Artemia*).

A young pair of popular Boesmani; *the male is the lower fish.*

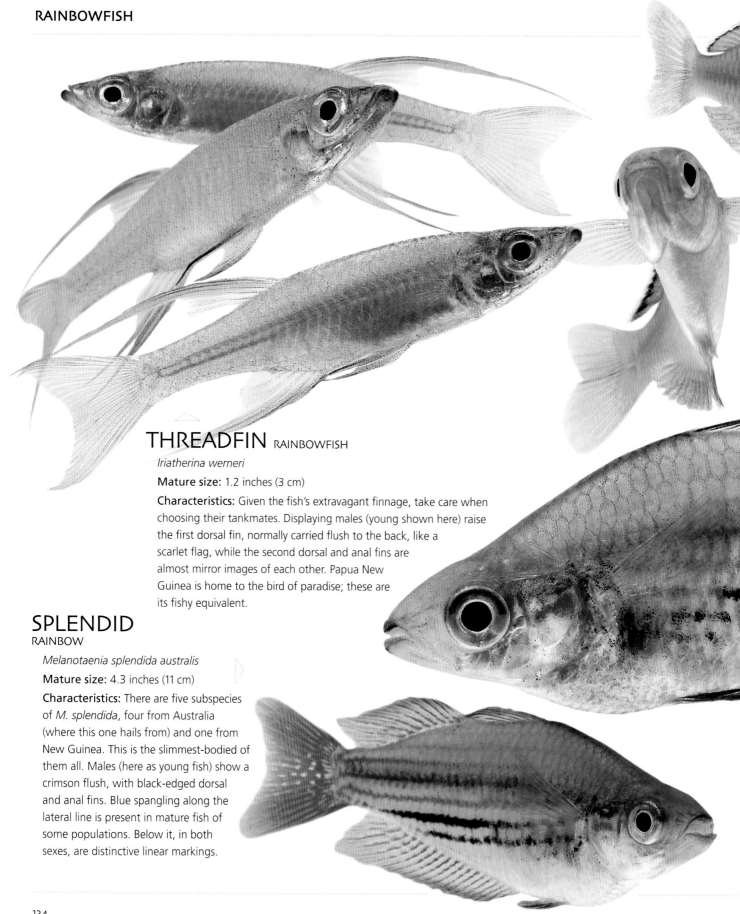

THREADFIN RAINBOWFISH

Iriatherina werneri

Mature size: 1.2 inches (3 cm)

Characteristics: Given the fish's extravagant finnage, take care when choosing their tankmates. Displaying males (young shown here) raise the first dorsal fin, normally carried flush to the back, like a scarlet flag, while the second dorsal and anal fins are almost mirror images of each other. Papua New Guinea is home to the bird of paradise; these are its fishy equivalent.

SPLENDID RAINBOW

Melanotaenia splendida australis

Mature size: 4.3 inches (11 cm)

Characteristics: There are five subspecies of *M. splendida*, four from Australia (where this one hails from) and one from New Guinea. This is the slimmest-bodied of them all. Males (here as young fish) show a crimson flush, with black-edged dorsal and anal fins. Blue spangling along the lateral line is present in mature fish of some populations. Below it, in both sexes, are distinctive linear markings.

BOESMANI

Melanotaenia boesmani

Mature size: 3.5 inches (9 cm)

Characteristics: As males mature, their coloration fore and aft of the first dorsal fin is very distinctive. The head end flushes iridescent blue, silver and green, while the rear of the fish turns deep orange. Once they are in the mood, spawnings can last several days.

♂

♀

CELEBES
RAINBOWFISH

Marosatherina ladigesi

Mature size: 2.75 inches (7 cm)

Characteristics: The Celebes Rainbowfish, from Sulawesi, belongs to a family with several representatives from brackish waters. The blue flank reminds us of the Neon Tetra, and the finnage of mature males is striking — rays of the second dorsal and anal fins are elongated and separated. Females are drabber, but without their presence in the tank you will not see males display to their best advantage. These are young fish.

♀

♂

SALMON RED
RAINBOW

Glossolepis incisus

Mature size: 4.7 inches (12 cm)

Characteristics: Rainbowfish have a lot going for them — they are peaceable, adaptable, not fussy about their food and fairly easy to breed if you give them a spawning mop or a thickly planted tank. In this New Guinea species, the males (shown here) become quite deep-bodied as they mature and their coloration intensifies when the ladies are around.

LOACHES

*Clown prince —
Botia macracantha
with its motley markings.*

streamlining helps the fish hug the riverbed.

If you are looking for a peaceful loach, *Botia macracantha* is hard to beat. Droll, dramatically banded and gregarious, it is usually imported small, but can eventually attain 12 inches (30 cm). Aquarium spawnings are extremely rare; it is thought that breeding is triggered by the arrival of the rains.

Some *Botias* are quite sturdy and high-backed (*Botia morleti*, Hora's Loach), while others are slim and elongated. Taking matters to extremes, though, are the worm-like Kuhli Loaches (*Pangio* spp.), which for some reason are always recommended as community fish. But they are so secretive that once they are in the tank you rarely catch more than fleeting glimpses of them — and since they are more agile than a greased eel, it is virtually impossible ever to net them out again.

It's black, it's red-tailed, but the Red-tailed Black Shark is a shark in name only.

"Loach" is a catch-all name for two families of aquarium fish — Balitoridae (river and hillstream loaches) and the far more popular Cobitidae (including the familiar Clown Loach, belonging to the genus *Botia*). Members of the latter tend to have a flat belly profile and an aquadynamic snout adorned with prominent barbels, which are used to root out invertebrate food items. When facing into a current, this

*The Hillstream Loach
(Gastromyzon sp.) is
ideally shaped for its
fast-flowing habitat.*

& SHARKS

family that bear a superficial resemblance to the marine predators in general body shape and, in particular, the triangular dorsal fin. This illusion is most pronounced in the Red-tailed Black Shark *(Epalzeorhynchos bicolor),* because the velvet-black of the body extends into the dorsal fin. This, and the clearly delineated red caudal fin, give it a guaranteed place in the community aquarium, although, like other members of its genus, it can be grumpy and may pick on smaller individuals.

Many similar-looking banded loaches are regularly imported with only sketchy identification. Some of these can be very territorial, and it is definitely a case of keeping one per tank — and then only if there is plenty of cover for them in the form of rocks and driftwood. To avoid conflict in the aquarium, steer clear of any loach that is marked like a tiger or a zebra.

"Sharks" are not really sharks at all, but Asian members of the carp

are common targets, possibly because they are mistaken for other sharks. Little physical damage is inflicted, but once a Red-tail gets a complex about any of its tankmates it is a case of more or less constant harassment.

No such bullying stains the character of the two Flying Foxes. *Crossocheilus siamensis* (Siamese Flying Fox) is more concerned with eating algae than fighting, and the lookalike *Epalzeorhynchos kalopterus* (Flying Fox) makes only small inroads into the greenery and is an inoffensive enough fish.

Tiger Barbs

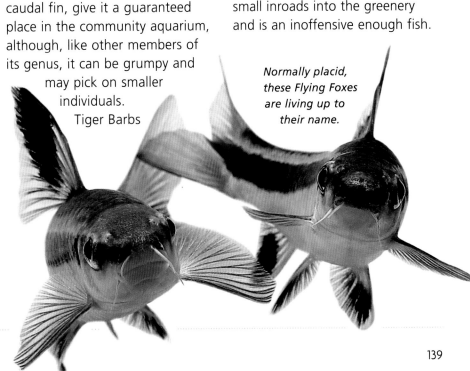

Normally placid, these Flying Foxes are living up to their name.

RAINBOW SHARK

Epalzeorhynchos frenatus

Mature size: 6 inches (15 cm)

Characteristics: This slimline shark, also known as the Ruby Shark, is an active midwater swimmer, and more peaceable than the RTBS. All the fins are orange-red (brighter in males) with a darker diamond-shaped blotch on the wrist of the tail. Dark barbels stand out beside a characteristic facial stripe. They are spawned commercially in ponds in Thailand, but aquarium breeding success is rare.

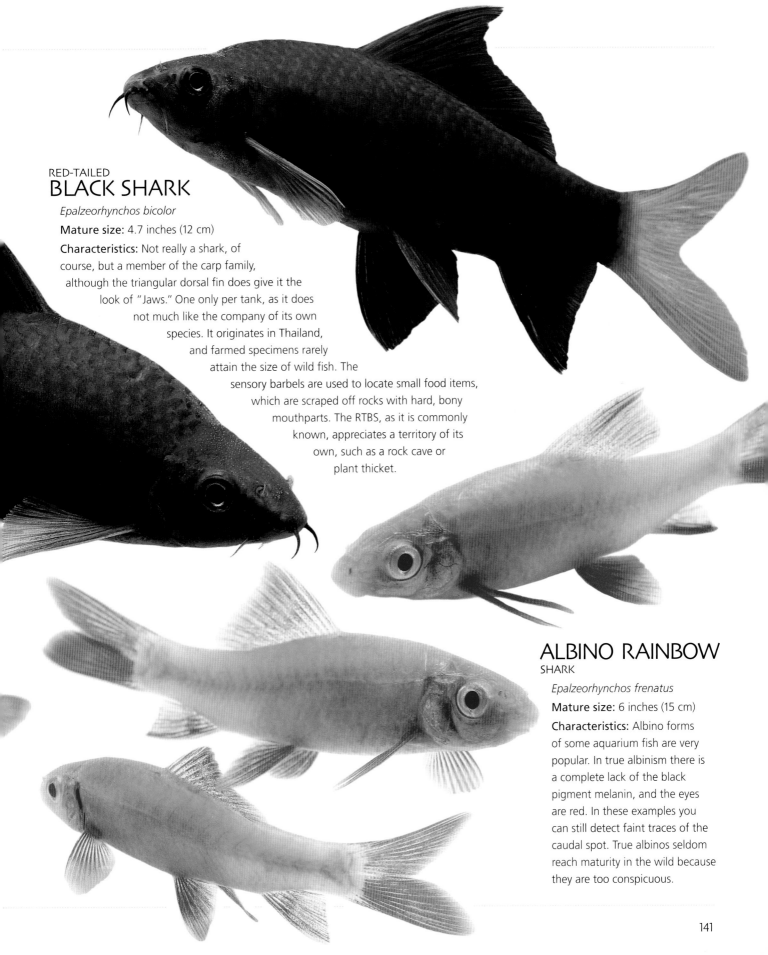

RED-TAILED
BLACK SHARK

Epalzeorhynchos bicolor

Mature size: 4.7 inches (12 cm)

Characteristics: Not really a shark, of course, but a member of the carp family, although the triangular dorsal fin does give it the look of "Jaws." One only per tank, as it does not much like the company of its own species. It originates in Thailand, and farmed specimens rarely attain the size of wild fish. The sensory barbels are used to locate small food items, which are scraped off rocks with hard, bony mouthparts. The RTBS, as it is commonly known, appreciates a territory of its own, such as a rock cave or plant thicket.

ALBINO RAINBOW
SHARK

Epalzeorhynchos frenatus

Mature size: 6 inches (15 cm)

Characteristics: Albino forms of some aquarium fish are very popular. In true albinism there is a complete lack of the black pigment melanin, and the eyes are red. In these examples you can still detect faint traces of the caudal spot. True albinos seldom reach maturity in the wild because they are too conspicuous.

BALA SHARK

Balantiocheilus melanopterus

Mature size: 14 inches (35 cm)

Characteristics: These sporty-looking silver fish with lemon-yellow-and-black finnage are extremely active members of the carp family, and look best in a shoal. They are great jumpers, so cover their tank closely. Aquarium spawnings are rare, but Bala Sharks are bred commercially in the Far East by pituitary injection, which takes pressure off the wild population. They are peaceable fish that prefer some vegetable matter in their diet.

CLOWN LOACH

Botia macracantha

Mature size: 12 inches (30 cm)

Characteristics: Nobody knows quite why Clown Loaches have the odd habit of lying on their sides for hours on end, but it seems to be quite normal. Fish bought in aquatic stores are usually babies or juveniles, and few people realize how large they can grow, given enough space. Their droll banded pattern, streamlined yet sturdy bodies and relatively peaceful nature make them popular aquarium subjects. However, they can be very sensitive to tank medications.

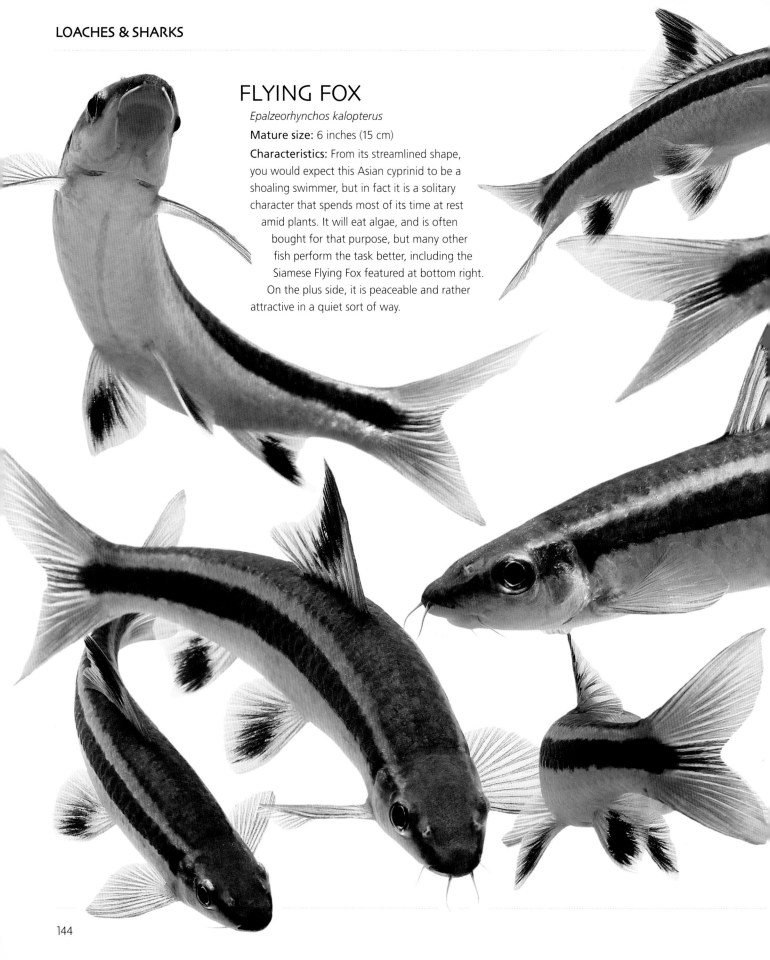

FLYING FOX

Epalzeorhynchos kalopterus

Mature size: 6 inches (15 cm)

Characteristics: From its streamlined shape, you would expect this Asian cyprinid to be a shoaling swimmer, but in fact it is a solitary character that spends most of its time at rest amid plants. It will eat algae, and is often bought for that purpose, but many other fish perform the task better, including the Siamese Flying Fox featured at bottom right. On the plus side, it is peaceable and rather attractive in a quiet sort of way.

SIAMESE FLYING FOX

Crossocheilus siamensis

Mature size: 5.5 inches (14 cm)

Characteristics: This fish is often confused with the other Flying Fox, even by importers, but if thread algae is a problem in your aquarium it is well worth tracking down. A fail-safe ID point is the absence of barbels. This is altogether a drabber and more thickset fish that earns its place by virtue of what it does, rather than what it looks like.

GOLD ZEBRA LOACH

Botia histrionica

Mature size: 2.5 inches (6.5 cm)

Characteristics: Formerly *Botia dario*, the Gold Zebra (or Bengal) Loach from India is intensively tank-bred in Far Eastern fish farms, but rarely spawned in the aquarium.

POLKADOT LOACH

Botia rostrata

Mature size: 2.4 inches (6 cm)

Characteristics: This moderately belligerent fish from India and Burma is also known as the Ladder Loach. Females are larger and have fewer light blotches — not that these fish have been successfully spawned in aquariums.

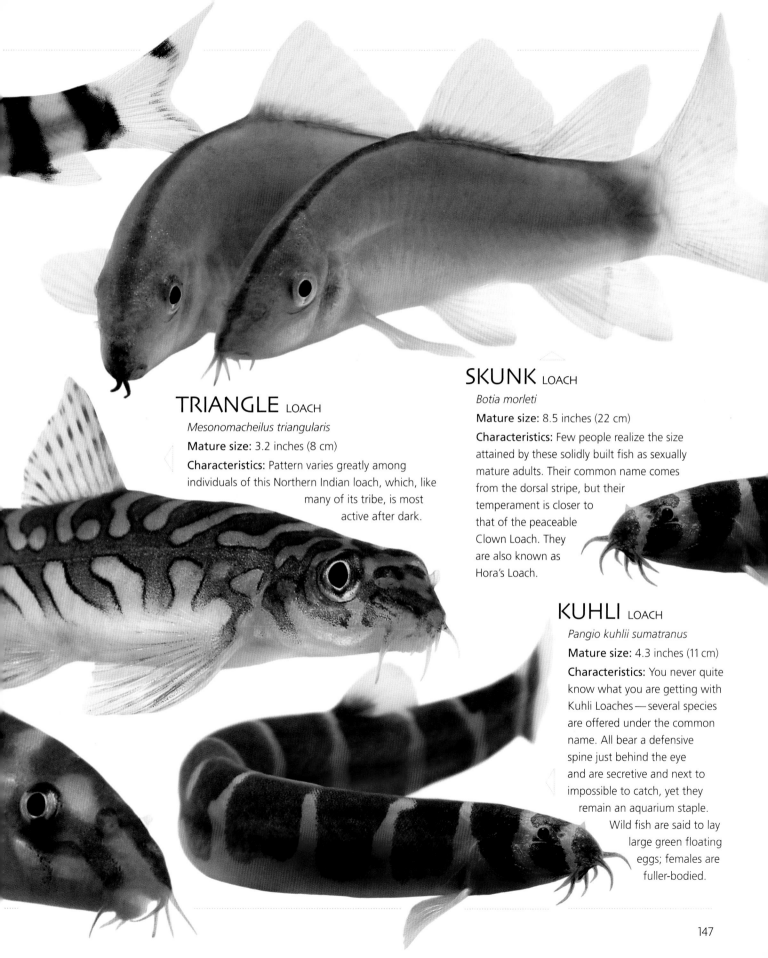

TRIANGLE LOACH

Mesonomacheilus triangularis

Mature size: 3.2 inches (8 cm)

Characteristics: Pattern varies greatly among individuals of this Northern Indian loach, which, like many of its tribe, is most active after dark.

SKUNK LOACH

Botia morleti

Mature size: 8.5 inches (22 cm)

Characteristics: Few people realize the size attained by these solidly built fish as sexually mature adults. Their common name comes from the dorsal stripe, but their temperament is closer to that of the peaceable Clown Loach. They are also known as Hora's Loach.

KUHLI LOACH

Pangio kuhlii sumatranus

Mature size: 4.3 inches (11 cm)

Characteristics: You never quite know what you are getting with Kuhli Loaches — several species are offered under the common name. All bear a defensive spine just behind the eye and are secretive and next to impossible to catch, yet they remain an aquarium staple. Wild fish are said to lay large green floating eggs; females are fuller-bodied.

RED TORPEDO

Crossocheilus denisonii

Mature size: 6 inches (15 cm)

Characteristics: By far the most striking member of its genus, this Indian fish sports a crimson pennant marking that runs along the flank from eye to dorsal fin, and echoes of the other body coloration in the deeply forked tail. It is not often seen in the hobby, which is a shame, because it has no vices, eats the usual fare and will not pick fights with its tankmates.

BLACK SHARK

Labeo chrysophekadion

Mature size: 24 inches (60 cm)

Characteristics: Young specimens of this most sharklike of sharks are a funereal matt black, but as they mature they develop an underlying bronze tint. They are grumpy rather than overtly aggressive. Usually bought small, they quickly outstrip other fish in a community tank, to which they are not really suited. Black Sharks eat plenty of green matter, so feed them algae wafers plus flake and freeze-dried insect foods.

BARBS

In basic outline, barbs are everyone's idea of a typical fish. These members of the worldwide carp family Cyprinidae all show the typical streamlined shape of midwater shoaling fish, and their lifestyle does not require bizarre adaptations to deal with a challenging habitat.

The term "barb" is unscientific, but we understand it to mean fish of the genus *Barbus* from both Africa and Asia (the old generic name *Puntius* is still seen in the aquarium trade). One or two other examples, such as the Tinfoil Barb *(Barbodes schwanenfeldi)* and Hard-lipped Barb *(Osteochilus hasselti)* sneak into this category too, but these are large fish that often end up being re-homed in public aquariums when they outgrow their quarters. Most barbs do not present us with such a problem.

Barbs feed on small aquatic insects and crustaceans and some plant matter. There are no predators to worry about in this group, although some species can

A burnished Copper Rosy Barb (Barbus conchonius).

A rather surprised-looking African Striped Barb (Barbus lineatus).

be boisterous when they are kept with long-finned tankmates. Tiger Barbs *(Barbus tetrazona tetrazona)* have the reputation of being fin-nippers, but in shoals of six or more this tendency is toned down.

With their bright coloration and iridescent scales flashing in the light as a shoal twists and turns, Asian barbs are aquarium naturals. Their common names often give clues as to their markings — Spanner, Black Ruby, Checker

Leptobarbus hoevenii —
*a largish barblike fish
from Southeast Asia.
The inset view shows
the fine barbels.*

and Five-banded Barbs. While Asian barbs are vibrant, African species are subtler propositions. The small, exquisite *Barbus jae* is rose-tinted with neat lateral banding, while the Butterfly Barb *(Barbus hulstaerti)* best shows off its contrasting black, silver and yellow markings against a dark tank background.

Long-finned variant barbs of the common aquarium species — such as the Copper Rosy Barb — are the result of commercial breeding, and one of them — the Schubert's, or Golden Barb — is unknown in the wild. That adds a touch of mystery to these otherwise uncomplicated fish.

Barbs are egg-scatterers, showing no parental care — as potential prey, the adults cannot stay around to look after their broods, so they play a percentage game. Only a very few eggs, adhering to fine-leaved plants or resting in the substrate, survive to maturity.

However, it is habitat deprivation, not natural predation, that threatens some species in the wild. The Cherry Barb *(Barbus titteya)* of Sri Lanka hangs on, thanks to conservation schemes that can easily fail during periods of political unrest, yet it is bred by the million in Far Eastern fish farms.

*Exuberant Tiger Barbs —
their charm offensive
easily overcomes a
reputation for
fin-nipping.*

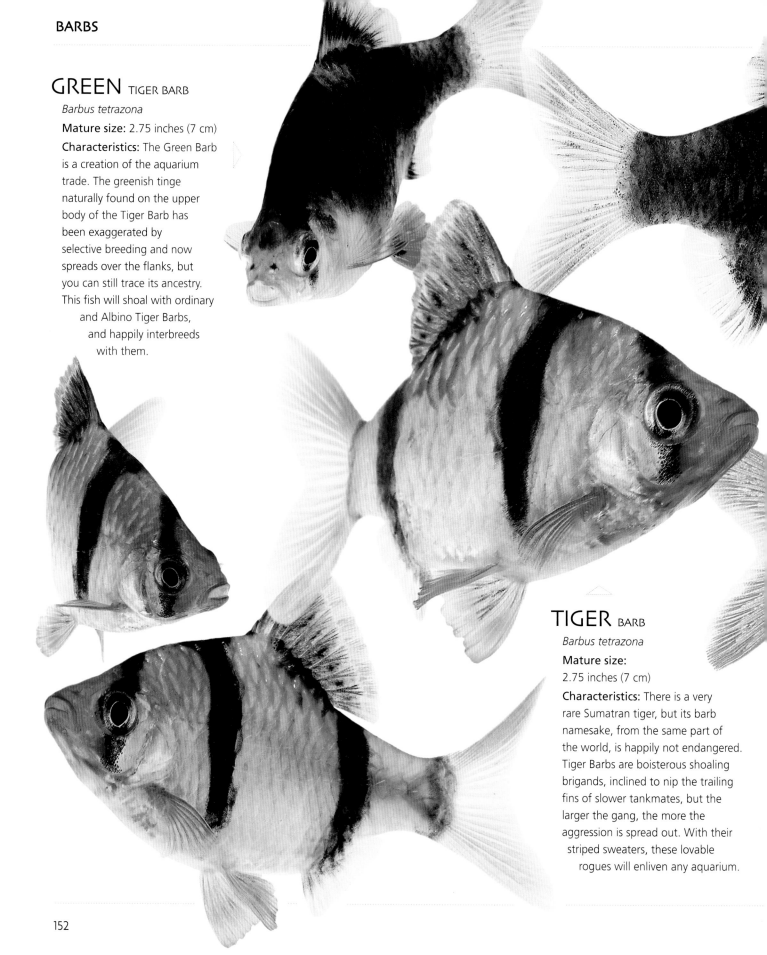

GREEN TIGER BARB

Barbus tetrazona

Mature size: 2.75 inches (7 cm)

Characteristics: The Green Barb is a creation of the aquarium trade. The greenish tinge naturally found on the upper body of the Tiger Barb has been exaggerated by selective breeding and now spreads over the flanks, but you can still trace its ancestry. This fish will shoal with ordinary and Albino Tiger Barbs, and happily interbreeds with them.

TIGER BARB

Barbus tetrazona

Mature size:
2.75 inches (7 cm)

Characteristics: There is a very rare Sumatran tiger, but its barb namesake, from the same part of the world, is happily not endangered. Tiger Barbs are boisterous shoaling brigands, inclined to nip the trailing fins of slower tankmates, but the larger the gang, the more the aggression is spread out. With their striped sweaters, these lovable rogues will enliven any aquarium.

ALBINO TIGER BARB

Barbus tetrazona

Mature size: 2.75 inches (7 cm)

Characteristics: These are not true albinos, but a leucistic form of the Tiger Barb in which the eyes remain black. For some reason, these fish are not as aggressive as their naturally pigmented cousins.

COPPER
ROSY BARB

♀

♂

Barbus conchonius

Mature size: 6 inches (15 cm)

Characteristics: Long a popular aquarium fish, the Rosy Barb is named for the reddish tinge that suffuses the silvery flanks and intensifies when pairs are ready to spawn, which they do by scattering eggs among plants. This farmed variant displays elongated finnage, which in some company can turn it from a potential fin-worrier into a victim itself. Note the gold-edged black spot forward of the tail. These undemanding fish will eat almost anything, but appreciate some live food. They shoal from midwater downward.

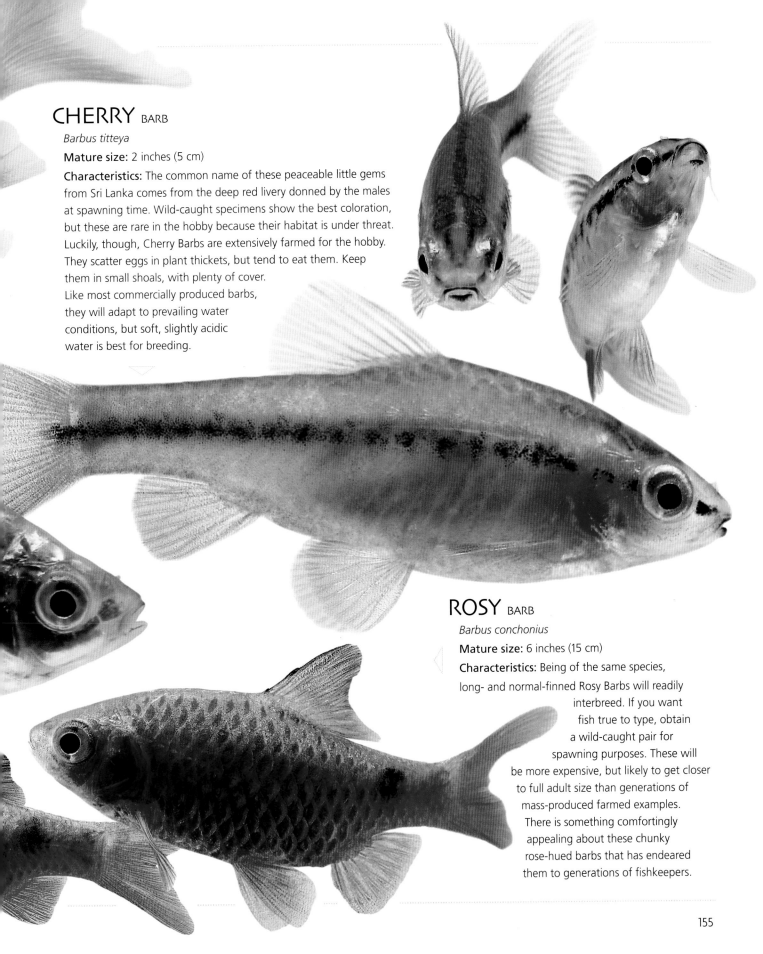

CHERRY BARB

Barbus titteya

Mature size: 2 inches (5 cm)

Characteristics: The common name of these peaceable little gems from Sri Lanka comes from the deep red livery donned by the males at spawning time. Wild-caught specimens show the best coloration, but these are rare in the hobby because their habitat is under threat. Luckily, though, Cherry Barbs are extensively farmed for the hobby. They scatter eggs in plant thickets, but tend to eat them. Keep them in small shoals, with plenty of cover. Like most commercially produced barbs, they will adapt to prevailing water conditions, but soft, slightly acidic water is best for breeding.

ROSY BARB

Barbus conchonius

Mature size: 6 inches (15 cm)

Characteristics: Being of the same species, long- and normal-finned Rosy Barbs will readily interbreed. If you want fish true to type, obtain a wild-caught pair for spawning purposes. These will be more expensive, but likely to get closer to full adult size than generations of mass-produced farmed examples. There is something comfortingly appealing about these chunky rose-hued barbs that has endeared them to generations of fishkeepers.

STRIPED BARB

Barbus lineatus

Mature size: 4.7 inches (12 cm)

Characteristics: Males of this barb from Africa's Zaire Basin are slim and darkly pin-striped, while females are less distinctly marked and much more rotund in their gravid state. This is quite an easy barb to spawn in plant thickets, and fry are numbered in the thousands. Note the large eyes and long, whiskery barbels.

RED-FINNED
CIGAR SHARK

Leptobarbus hoevenii

Mature size: 20 inches (50 cm)

Characteristics: This is a rewarding Southeast Asian fish for large community tanks but, as an active midwater swimmer, it requires plenty of space to realize its full growth potential. The red deeply forked, rounded tail lobes, streamlined shape and lips accented in black are points to appreciate in an otherwise subtle fish. The sexes are indistinguishable.

ARULIUS BARB

Barbus arulius

Mature size: 4.7 inches (12 cm)

Characteristics: The bigger this Indian barb grows, the better it looks — males develop very elongated dorsal fin rays, and both sexes put on a show of subtle body coloration shot through with green, blue and violet, barred in black. This robust and thickset fish is easy to spawn and unfussy about what it eats, although live insect food is always welcome.

Enter the weird and wonderful world of the oddballs. You would be pushed to invent them in your wildest fantasies, but however strange they seem, nothing in nature happens by accident.

ODDBA

What qualifies a fish as an oddball? On lifestyle alone, the scope is limitless; what are we to make, for instance, of the Lake Malawi cichlid *Nimbochromis polystigma*? It lies on its side, looking just like a dead and decaying fish, only to spring to life and engulf bystanders — a case of "prey" turned predator.

Yet that candidate somehow does not qualify. Fishkeepers tend to make snap visual judgments, and oddballs challenge our concept of what normal fish should look like. Consider the Pufferfish *(Tetraodon biocellatus)*, wide-eyed and cute. Yet parts of the fish contain a deadly nerve toxin. Threaten it and it inflates with water to a size most predators cannot handle.

Many other fish carry bizarre weaponry. Freshwater Stingrays *(Potamotrygon* spp.), like their marine counterparts, have barbed, poisonous spines in their whiplike tails, and we have all heard of the

Nature's own inflatable — a freshwater Pufferfish.

Rare predator — the magnificent Arowana, for large tanks only.

legendary jolts delivered by the Electric Eel. There are lower-voltage oddballs too, such as the Elephantnose *(Gnathonemus petersi)* from the Zaire Basin. The bizarre appearance of its

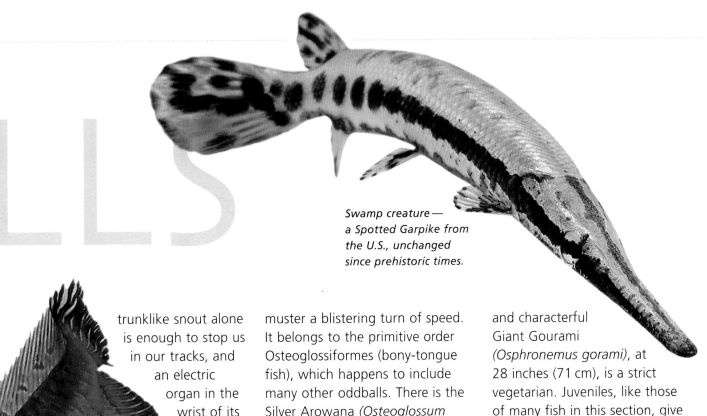

*Swamp creature —
a Spotted Garpike from
the U.S., unchanged
since prehistoric times.*

trunklike snout alone is enough to stop us in our tracks, and an electric organ in the wrist of its tail has the same effect on territorial rivals. Cast an eye over the Bichirs (*Polypterus* spp.) and their relative the eel-like Ropefish (*Erpetoichthys calabaricus*). Real oddities these, as they perch on their ventral fins waiting for prey, looking like creatures from a primeval swamp.

How about the African Butterflyfish *(Pantodon buchholzi)*? Its extravagant, filamentous fins spread wide, it hangs motionless for hours under the surface. Yet when prey approaches, it can

muster a blistering turn of speed. It belongs to the primitive order Osteoglossiformes (bony-tongue fish), which happens to include many other oddballs. There is the Silver Arowana *(Osteoglossum bicirrhosum)* from Amazonia, with its huge, bladelike body and forward-jutting barbels; the world's largest freshwater fish, the massive Piracuru *(Arapaima gigas)*; and the African and Asian Knifefish, propelled along by waves from an anal fin that runs almost the full length of the body.

Many larger oddballs are carnivores, but the supremely ugly

and characterful Giant Gourami *(Osphronemus gorami)*, at 28 inches (71 cm), is a strict vegetarian. Juveniles, like those of many fish in this section, give few clues to their eventual size or adult appearance, so beware of impulse buying.

For reasons of size, temperament, feeding and water requirements, some oddballs are best left to the specialist fishkeeper, but anyone with an eye for the bizarre will still find plenty of suitable material at their aquatic store.

*Baleful beast —
snakeheads are
bad news for
anything smaller
that swims.*

BUTTERFLYFISH

Pantodon buchholzi

Mature size: 4 inches (10 cm)

Characteristics: Surely flying fish are found only in the oceans? Then what can we make of this oddity, pectoral fins spread like the wings of an early monoplane? The paired ventral fins are equally strange, each divided into four long filaments. So still does this fish lie that it could be a mirage reflected in the surface mirror. It is a different story at night, when this predator wakes and seeks out small fish or aquatic insects. The Butterflyfish is found in forest pools in tropical West Africa. When startled, it can leap from the water and glide several yards, so obviously it needs a large tank with a close-fitting lid. It is almost impossible to wean *Pantodon* off live food items, but as long as its tankmates are too large to be engulfed by its large, upturned mouth, it is a peaceable fish to keep. Up to 400 large, floating eggs are laid, but reports of captive breeding are rare.

SILVER AROWANA

Osteoglossum bicirrhosum

Mature size: 48 inches (120 cm)

Characteristics: Arowanas look prehistoric, and are indeed members of an ancient order of fish. When the mouth of this bony-tongue fish from Amazonia is closed, the head appears compact and wedge-shaped. But when the sensitive barbels (below) detect the movement of insects on the surface, the jaws open wide to engulf them. As a prized occupant of a very large aquarium, either on its own or with tankmates too large to swallow, it is in fact quite peaceful. Imported

specimens are most likely to have been farmed in the Far East. In the wild, females lay large, floating eggs; on hatching, the baby Arowana carries a huge yolk sac around for some weeks and does not need to feed until this has been absorbed.

SNAKEHEAD

Channa asiatica

Mature size: 12 inches (30 cm)

Characteristics: A southern Asian "pet" fish of some charm, and interesting for the male's habit of guarding clutches of floating eggs. All Snakeheads possess a labyrinth organ for breathing atmospheric air, and are very tenacious of life. The Snakehead is best kept on its own in a large specimen tank, as it is a piscivore through and through.

RED SNAKEHEAD

Channa micropeltes

Mature size: 39 inches (100 cm)

Characteristics: The Red Snakehead, often imported as a cute striped baby only a few inches long, is the ultimate catch-out for aquarists. As it grows, its coloration and pattern become more subdued, but there is no denying that a mature specimen, which will soon come to recognize its owner, is an aquarium talking point.

BLACK GHOST KNIFEFISH

Apteronotus albifrons

Mature size: 20 inches (50 cm)

Characteristics: South American Indians fear these fish as the spirits of drowned fishermen, and the Black Ghost Knifefish is indeed weird. Just look at the doglike face, the banded tail, which looks as though it belongs to another creature, and the massive anal fin that propels the fish backward or forward with a wavelike motion. A close relative is the Electric Eel, but the rather weak field generated by the Black Ghost is used only for navigation.

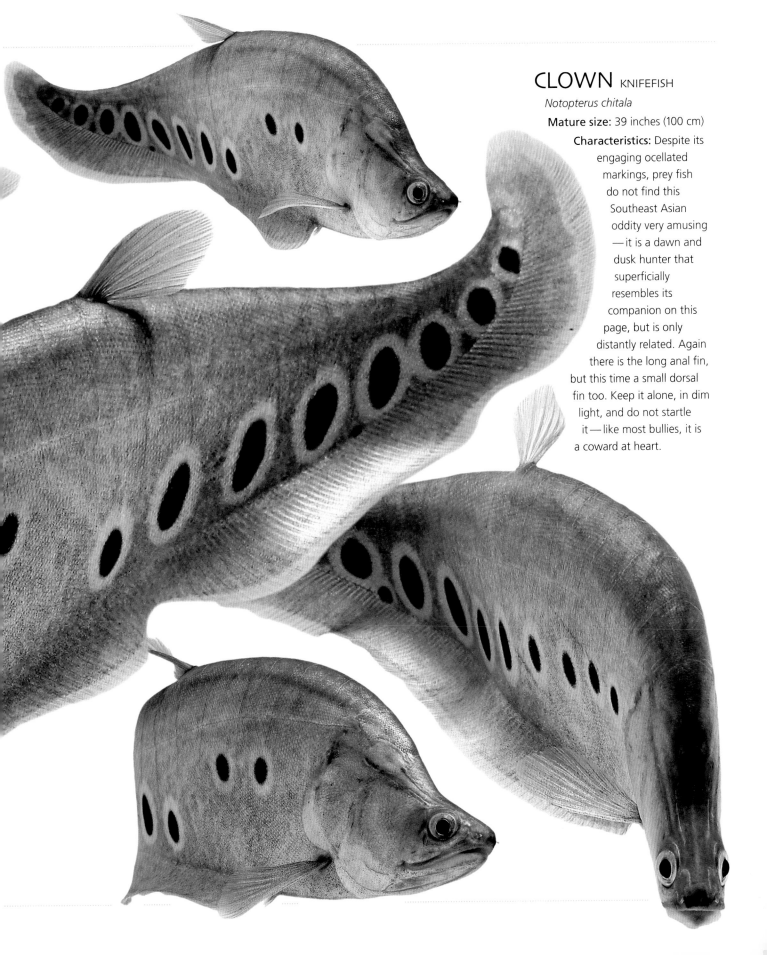

CLOWN KNIFEFISH

Notopterus chitala

Mature size: 39 inches (100 cm)

Characteristics: Despite its engaging ocellated markings, prey fish do not find this Southeast Asian oddity very amusing —it is a dawn and dusk hunter that superficially resembles its companion on this page, but is only distantly related. Again there is the long anal fin, but this time a small dorsal fin too. Keep it alone, in dim light, and do not startle it—like most bullies, it is a coward at heart.

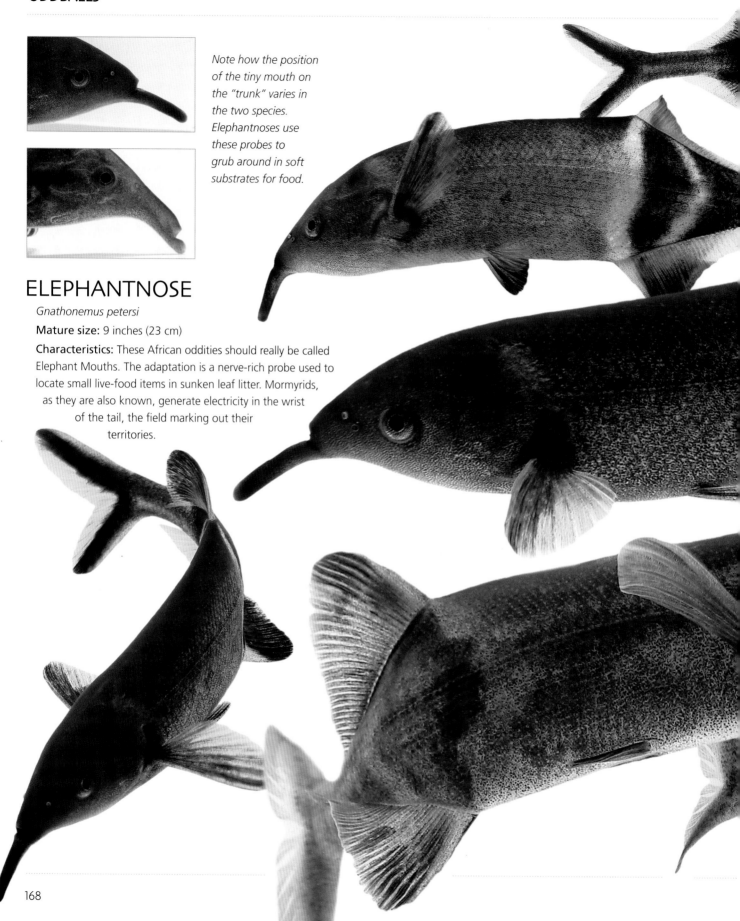

Note how the position of the tiny mouth on the "trunk" varies in the two species. Elephantnoses use these probes to grub around in soft substrates for food.

ELEPHANTNOSE

Gnathonemus petersi

Mature size: 9 inches (23 cm)

Characteristics: These African oddities should really be called Elephant Mouths. The adaptation is a nerve-rich probe used to locate small live-food items in sunken leaf litter. Mormyrids, as they are also known, generate electricity in the wrist of the tail, the field marking out their territories.

MARBLED ELEPHANTNOSE
Campylomormyrus tamandua

Mature size: 17 inches (43 cm)

Characteristics: In the Marbled Elephantnose, the upper and lower jaws form a trunklike mouth. In their African homeland, breeding is triggered by the rains, which swell the rivers and provide food for the youngsters. Elephantnoses are shy creatures and best kept only with their own kind.

ORNATE BICHIR

Polypterus ornatipinnis

Mature size: 12 inches (30 cm)

Characteristics: Glossy scales like protective body plates; an eel-like body; not one dorsal fin but many, held erect like tiny flags. All this, plus tube nostrils and pectoral fins, which the creature seems to want to use more like legs, makes the Bichir a real oddity of the fish world, and that is why it is sometimes kept. But beware — it is a predator, and its tankmates need to be big and robust. It hunts by night, locating its prey with tube nostrils (inset) that are capable of independent movement.

DELHEZI'S BICHIR

Polypterus delhezi

Mature size: 14 inches (35 cm)

Characteristics: From Zaire, Africa, comes one of the more attractive Bichirs. Like other members of the genus, it uses an adapted swim bladder to breathe atmospheric air and needs to surface every now and again to take a gulp. Otherwise, it spends most of its time resting on the bottom, looking unnervingly intelligent, as if waiting for an evolutionary quantum leap to pitch it onto land. Its scales are covered with a hard, lacquerlike coating.

SPOTTED GARPIKE

Lepisosteus oculatus

Mature size: 48 inches (120 cm)

Characteristics: A survivor from the Jurassic period (about 150–200 million years ago), this U.S. citizen is a lurk predator that can really only be kept with its own kind in a large, planted tank. It will, however, learn to take dead food. With its shiny coat of primitive ganoid scales, it has a rather forbidding sharp-nosed prehistoric charm — a modern ichthyosaur.

ROPEFISH

Erpetoichthys calabaricus

Mature size: 16 inches (40 cm)

Characteristics: This is the stretch limo among polypterids and, like the eel it rather resembles, a great escape artist. The tube nostrils featured below are sensory organs that perform the same function as the barbels of catfish. Hard protective scales still allow huge flexibility, enabling the Ropefish to squeeze into quite small crevices or even burrow into the substrate. Although a predator, it is a danger only to bite-sized fish, and is peaceable with other members of its own species.

Starting with the Comet Goldfish, breeding by man has produced fancy goldfish in a wide diversity of patterns, body forms and finnage. These cold-water fish are either beautiful or bizarre. You decide.

Gaze into a tank of fancy goldfish and be amazed at their diversity of form and coloration. Here a matt-black Moor, trailing resplendent finnage and looking out on the world through boggling telescopic eyes; there a Ranchu, its head adorned with a hood looking like a succulent raspberry. And resting on the bottom is a Calico Bubble-eye, so bizarre you would not imagine such a creature could exist, had you not seen it for real. These weird and wonderful creatures are all members of a single species, *Carassius auratus.*

The Chinese, and later the Japanese and Europeans, have been culturing fancy goldfish for centuries. The raw material is an innocuous little brown carp, a fish remarkable only for its hardiness. The "fairground" goldfish is one stage removed from the ancestral type, a simple mutation of the skin into a bright golden orange. These common goldfish are quite able to survive in a garden pond. Many of the varieties shown on the following pages, however, are not suited to the great outdoors; they are highly valued aquarium fish that richly deserve the pampering of their owners.

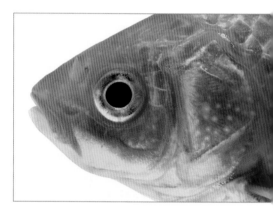

Males in breeding condition carry spawning tubercles on their gill plates.

Think of any anatomical characteristic, and chances are that fancy goldfish will carry them to extremes. Eyes, body shape, finnage and coloration have all been changed. There are matt and metallic goldfish in single and multi-patterned

Fantails are prime examples of twin-tailed man-made goldfish.

ISH

A canary-yellow, rather than the usual orange, Comet Goldfish.

colorations. Some bear a tall, proud dorsal fin, while in others this fin is absent. Most fancies have twin tails, either short and stiff or flowing like a veil of delicate silk. Their palette makes the term "goldfish" redundant, when selective breeding has led to so many other hues. And what are we to make of pompon nostrils, extravagant hoods or a skin clothed in scales like the dimpling on a golf ball?

Extreme fins, eyes and body shape — a calico Telescope-eye.

In common with toy dogs or lop-eared rabbits, fancy goldfish are there to be admired, although some of us will always prefer creatures as nature intended, with form tied closely to function. But the need to swim fast from predators or to be drably clad for protection no longer applies to fish that spend their lives among their own peaceable kind in aquariums. We should not try to attribute human emotions to them and imagine that a Moor or a Celestial (so named because its eyes are directed upward to the heavens) is conscious of its appearance.

As long as basic needs are met — good food and impeccable water quality — fancy goldfish will lead a life of simple contentment and provide endless enjoyment and fascination for their admiring keepers.

COMET GOLDFISH

Mature size: 8 inches (20 cm)

Characteristics: This fish has set many hobbyists on the road to keeping more exotic, tropical species—a childhood pet that would survive everything until its owner developed better fishkeeping skills. The Comet Goldfish is closely related to the Crucian Carp *(Carassius carassius)* and will happily live in ponds, but first the Chinese, then the Japanese, worked on this chunky little carp to selectively breed fancy varieties. It is hard to appreciate that some of these are the same fish; the more extravagant the finnage, body shape, coloration and eye type, the less able these fancies are to live outdoors. Not so the world's most popular pet, the "common" goldfish.

SHUBUNKIN

Mature size: 6 inches (15 cm)

Characteristics: It is mainly their coloration that sets Shubunkins apart from Comet Goldfish — a suit of matt and nacreous scales in red, blue, orange, white and black. "Pedigree" London (short-finned) and Bristol (long-finned) Shubunkins are less common than these "everyday" varieties.

FANTAIL

Mature size: 8 inches (20 cm)

Characteristics: Fantails, the Western equivalent of the Ryukin, show an egg-shaped body, a tall dorsal fin and a double tail, carried quite stiffly. This highly variable variety is available in all shades, metallic or nacreous, and with or without telescope-eyes. The young fish to the left may lose the black pigment as they mature, although the way this carries on into the finnage suggests that in this instance it is a stable trait. Note the lack of a shoulder hump on all these examples. Fantails are tough, although prone to swim bladder trouble if chilled.

RED-AND-SILVER RYUKIN

Mature size: 8 inches (20 cm)

Characteristics: The very deep-bodied Ryukin is a Japanese strain of fancy goldfish named after the Ryukyu Islands between Japan and Taiwan. The back slopes up steeply from a somewhat pointed head, and the single dorsal fin is carried proudly, adding to the impression of height. The caudal fin typically has four lobes, although three are also acceptable. This variety has normal eyes and no hood, and swims well, despite the extravagant finnage. All the usual varieties, including calico, are represented, but red-and-silver is the most prized.

CALICO RYUKIN

"Calico" describes goldfish marked with red, black, blue, white and orange. They are available with matt or metallic scales. These Ryukin are typical of the type, every bit as bright as koi, but requiring much less space to keep.

ORANDA

Mature size: 8 inches (20 cm)

Characteristics: In Orandas, developed in Holland, the body is actually deeper than it is long. This can predispose the fish to swim bladder trouble, as that organ is compressed and foreshortened. The twin caudal fins resemble those of a Veiltail, and are ideally squared off, rather than forked, but the defining characteristic is the hood (or wen) composed of soft, wartlike growths. These take some time to develop, and in some examples never do, or only poorly. Typically the hood starts on the top of the head and spreads over the cheeks and around the eyes. Coloration is highly variable, and black pigment in young red fish (shown here in the middle) is usually lost over time. The baby metallic veiltail-type Orandas shown at right are only nine weeks old.

A still-developing Oranda hood. Newly emerging growths appear as white spots, not to be confused with the disease of that name.

RED-AND-SILVER
ORANDA

Mature size: 8 inches (20 cm)

Characteristics: These mature Red-and-Silver Orandas show exceptionally good dorsal and tail finnage, which balances the very deep body. Note how the hood of the fish at the bottom of the page is almost covering the eyes. You can see how two or three Orandas kept together will give an ever-changing perspective of form, depending on the direction in which they swim.

RED METALLIC ORANDA

Mature size: 8 inches (20 cm)

Characteristics: A red hood on a red fish shows the two shades that can result from the presence or absence of scales. On the metallic body, the coloration is toned down to more of an orange, while on the head the wen is more crimson—because here, of course, there are no scales to reflect back the light. The Japanese call red *hi* and in koi and goldfish they recognize many more shades than a Westerner could begin to imagine.

REDCAP ORANDA

Mature size: 8 inches (20 cm)

Characteristics: Looking like surprised Regency aristocrats in red fright wigs, these fish are an example of beauty in simplicity. The hood, unlike in other Orandas, is confined to the top of the head, where each warty element should be of equal size. The body should be silver, and the finnage (here of the veiltail-type) pure white, with no red intruding. Young fish with broad heads are preferred, as this gives the most room for the wen to develop.

TELESCOPE-EYE

Mature size: 4 inches (10 cm)

Characteristics: The overall body shape is similar to that of the Veiltail, with a single dorsal fin and pointed tips to the other fins, all of which should be paired. The eyes protrude from the head—hence the variety name. The caudal fin should be forked to about a quarter of its length and about three-quarters of the body length. In good specimens, the eyes should be at the tip of truncated cone-shaped protuberances and should be symmetrical. As the fish swims, the dorsal fin should be erect and the caudal fin should flow.

The Telescope-eye is also known as the Globe-eye in Britain and as the Dragon Fish in the Far East.

You can also find self (single) shade metallic, variegated or calico Telescope-eyes. The calicos must be bright, with a blue background and brown, orange, red, yellow and violet patches, spotted with black.

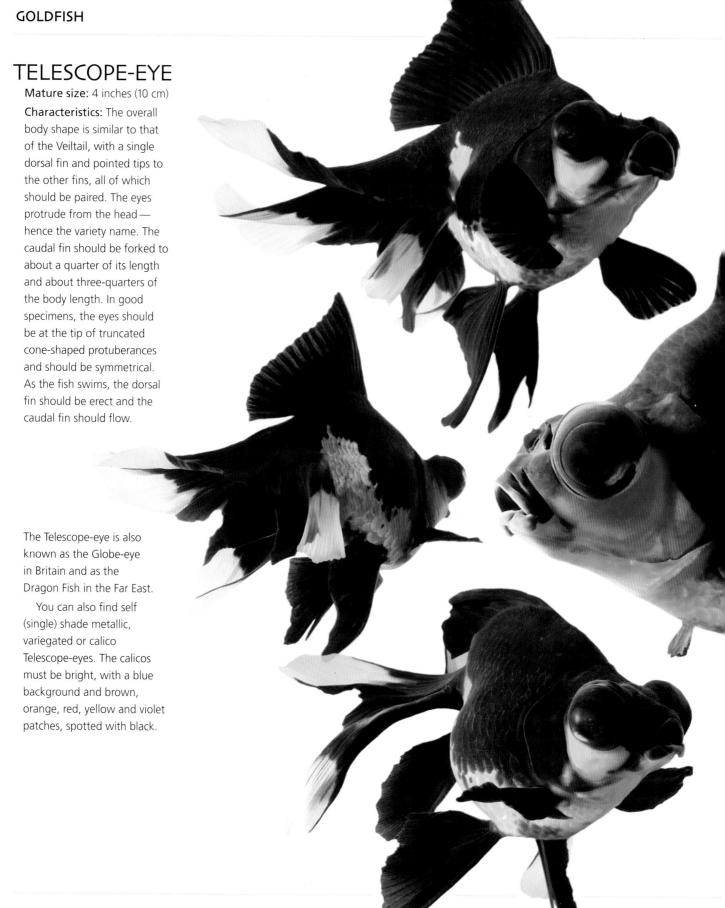

MOOR

Mature size: 6–10 inches (15–25 cm)

Characteristics: A Moor is a black Veiltail with telescope-eyes, and although perfect examples are very hard to find, the variety is enduringly popular. You would think that such a short-bodied fish would lack grace, but the flowing single dorsal and other, paired fins lend it an unlikely dignity, though it is not the fastest of swimmers. Ideally, Moors should be a uniform velvety black, but bronze scalation tends to show through with age.

All fancies of this general shape require a deep tank (20 inches/50 cm minimum). What is going on behind those unfathomable dark eyes? Goldfish are popularly thought of as having an attention span of seconds — if so, how come they recognize their owners?

PANDA BUTTERFLY

Mature size: 6–10 inches (15–25 cm)

Characteristics: Black-and-white chic in the animal kingdom is shared among magpies, skunks and, of course, giant pandas — after which this variant of the Moor is named. The silver-and-black pattern is highly variable, but when dark telescope-eyes appear on a white face the illusion is complete. Other markings need not necessarily be symmetrical, but should balance out. The Butterfly name refers to the tail which, when viewed from above, reminds one strongly of the wings of a swallowtail. The twin caudal fins should be fully divided and spread.

RED BUTTERFLY MOOR

Mature size: 6–10 inches (15–25 cm)

Characteristics: Moors are not always black,
which is why the old name "Black Moor" is still useful
in describing those that are, even though purists will not
use it. These fish, of course, are red, with only a touch of
silver around the telescope-eyes and below the chin, and
they share with the Panda the butterfly-shaped twin tail.
The tail is divided but not forked, in true Veiltail fashion.

BLACK
RANCHU

Mature size: 8 inches (20 cm)

Characteristics: They could be mistaken for giant aquatic loganberries, but these Black Ranchu are superb examples of fancy goldfish that are prized as highly as koi by the Japanese, who refined them from the very similar Chinese Lionhead. The main difference is that the Ranchu's back arches more steeply down toward the fully divided caudal fin which, as a result, is carried downward. Hood (wen) development is excellent in these matt-black beauties, extending over the top of the head and around the cheeks, almost (but not quite) covering the eyes. Ranchu are remarkably strong and hardy, and good swimmers.

RED-AND-SILVER METALLIC
RANCHU

Mature size: 8 inches (20 cm)

Characteristics: Ranchu are bred in metallic self-coloration or combinations, as well as calico. The Red-and-Silver type is popular, as this gives the best possible contrast between the hooded head and the body. These fish are still juveniles, yet to fully develop the wen. The top view shows just how the tail should be configured. Note the smooth dorsal profile, with no suggestion of a vestigial fin. Rigorous culling is needed to maintain the breed standards, but a good Ranchu is a sight to behold, giving an impression of restrained power.

These are nine-week-old Red Metallic Ranchu, in which the hoods are almost absent. Full development may take up to three years.

RED-AND-SILVER METALLIC
RANCHU

Mature size: 8 inches (20 cm)

Characteristics: Red-and-Silver Ranchu are not just
confined to the Sarasa type (red head, silver body); the twin
coloration, as here, can flow in a more informal pattern that
most resembles the markings on the Kohaku, a type of koi.
In the Jadehead, the face and hood are white and the rest of
the fish is red and silver. There are also examples where the
red extends along the back and is carried on into the
finnage. Indeed, new variants are appearing all the time.

RED METALLIC
RANCHU

Mature size: 8 inches (20 cm)

Characteristics: Sometimes simplicity of coloration is the hardest thing of all to accomplish in a fancy goldfish. The tendency of all fish is to have a dark back and a pale belly, the best camouflage against predators coming at them from above and below. Small wonder, then, that these Red Metallic Ranchu show areas of silver too. They are still fine fish, the steeply angled backs being especially well pronounced. Hood development is not as extreme as in the Black Ranchu on page 194.

CHOCOLATE
POMPON

Mature size: 8 inches (20 cm)

Characteristics: Goldfish have paired nostrils opening into a U-shaped tube lined with sensory cells that smell, and taste, the water. Each nostril is covered by a tiny flap of skin known as the nasal septum, but in Pompons these septa, or narial excrescences, have become exaggerated into fleshy balls that look rather like the tools of the cheerleader's trade. The bigger they are, the better — within reason. They should not be so large that they are sucked into the fish's mouth as it breathes. Pompons can be of the fantail-type, with a dorsal fin, or the lionhead-type, without. Good examples show a contrast in coloration between the nasal growths and the rest of the body. They occur in all the usual fancy goldfish hues.

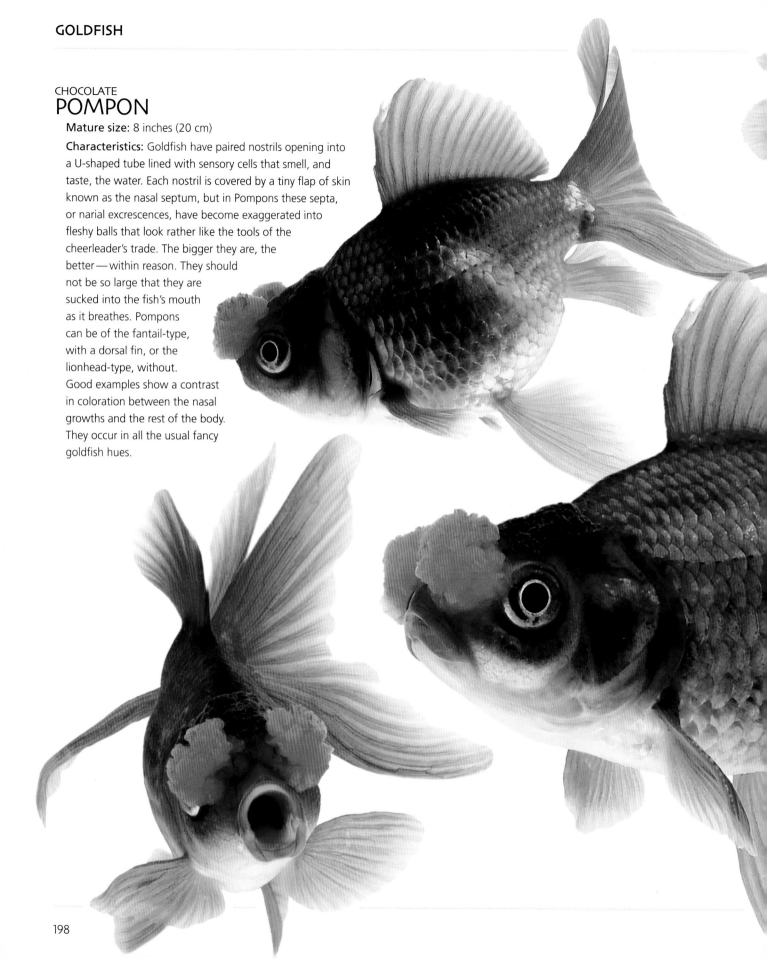

HAMA NISHIKI

Mature size: 8 inches (20 cm)

Characteristics: A Hama Nishiki, is actually a Pearlscale Oranda with a spherical body and slightly longer fins than the true Oranda. Like that more common fish, it has a hood (wen) that covers only the top of the pointed head. Some examples also have pompons, so it is a bit of an all-purpose fancy goldfish. It comes in several coloration patterns, metallic- and matt-scaled. Should one of the domed scales be rubbed off accidentally, it will grow back as a normal scale, one reason for keeping show fish only with their own, sedate kind.

CROWN
PEARLSCALE

Mature size: 6 inches (15 cm)

Characteristics: Breeders of fancy goldfish typically work on body shape, finnage, coloration and eye development. However, in the Pearlscale, another mutation involves the structure of the scales. Each is thickened in the center by a deposit of calcium carbonate (chalk), which makes it stand proud, so the overall effect is of the dimples on a golf ball — very visible on the Hama Nashiki (opposite). Other characteristics are a rounded body, forked and divided tail fin carried high, and a rather dainty head and small mouth. These examples also have an oranda-type hood. Calico Pearlscales are particularly attractive, and surprisingly hardy.

BRONZE BUBBLE-EYE

Mature size: 8 inches (20 cm)

Characteristics: Little beauties or wobbly weirdos? There is no neutral territory of opinion with Bubble-eyes. Coloration is secondary to those huge, fluid-filled pouches, so this variety can be found in self (single) shades—bronze, as here, red or yellow, or in various calico or metallic combinations. The pouches carry pigment, so they too can be of any coloration. In the paler types, blood vessels stand out on the bubbles, adding to the surreal effect. The twin tail should be deeply forked, and the dorsal fin is absent.

ORANGE
BUBBLE-EYE

Mature size: 8 inches (20 cm)

Characteristics: These extreme fancy goldfish are not great swimmers, as the eye pouches tend to wobble in motion and make their owners head-heavy, so their usual resting place is on the tank floor. Take care not to include any sharp gravel or decor, as those bubbles can easily be burst. In fact, this variety is best kept on its own, as it is not able to compete for food with more agile and better-sighted goldfish.

INDEX

Page numbers in **bold** refer to major entries, *italics* to captions in introductions and plain type to other text references.

A

"Aequidens"
  "A." coeruleopunctatus, **43**
  "A." pulcher, 31
African Blockhead, 9
African Glass Catfish, *105*
Albino Bronze Cory, **106**
Albino Rainbow Shark, **141**
Albino Tiger Barb, **153**
Altolamprologus
  compressiceps, **10–11**
Altum Angelfish, 31, **44**
Amblydoras hancockii, 105
American Flagfish, 79, **84**
Amphilophus citrinellus, 30, **32–33**
anabantids, 86–97
Ancistrus temminckii, **123**
Angel Catfish, **117**
Angelfish, 31, **45**
Anostomus
  A. anostomus, 53
  A. ternetzi, 67
Aphyosemion
  A. australe, 78, *79*
  A. ogoense pyrophore, **80**
  A. riggenbachi
   'Dibeng Gold,' **82**
Apistogramma, 31
  A. agassizii, **51**
  A. cacatuoides, **51**
  A. viejita, *30,* **50**
Apteronotus albifrons, **166**
Arapaima gigas, 159
Arowana, *158,* 159, **162–163**
Artemia, 133
Arulius Barb, **157**

Astronotus ocellatus, 31, **34–35**
Atheriformes, 132
Aulonocara, 8
  A. baenschi, **12**
  A. sp. 'Nagara,' **13**
  A. stuartgranti, 13

B

Baensch's Peacock, **12**
bagrids, 104
Balantiocheilus melanopterus, **142**
Bala Shark, **142**
Balitoridae, 138
Balloon Mollies, *69*
Banded Gourami, **95**
Banded Leporinus, *52, 53,* **67**
Barbodes schwanenfeldi, 150
barbs, 150–157. *See also Barbodes; Barbus*
  African, 151
  Asian, 150–151
Barbus
  B. arulius, **157**
  B. conchonius, *150,* 151, **154, 155**
  B. hulstaerti, 151
  B. jae, 151
  B. lineatus, *150,* **156**
  B. tetrazona, 150, *151,* **152–153**
  B. titteya, 151, **155**
Bedotia geayi, 132
Belontidae, 86
Bengal Danio, 99
Bengal Loach, **146**
Betta splendens (Betta), 71, 86, 87, **88–91**
Bichir, 159, **170–171**
Big Spot Hypostomus, **124**
Black Ghost Knifefish, **166**
Black Phantom Tetra, 52

Black Ranchu, **194**
Black Shark, **149**
Black Swordtail, **76**
Bleeding Heart Tetra, 52, **62**
Blue Acara, 31
Blue Diamond Discus, **49**
Blue Dwarf Gourami, **95**
Blue Jewel Cichlid, **17**
Blue One-spot Platy, **74**
Blue-point Acara, **43**
Blue Regal Peacock, **13**
Blue Turquoise Discus, **48**
Boesmani, *133,* **135**
bony-tongue fish, 159, 162–163
Botia
  B. dario, **146**
  B. histrionica, **146**
  B. macracantha, 138, **143**
  B. morleti, 138, **147**
  B. rostrata, **146**
Boulengerochromis microlepis, 9
Brachydanio, 98–99
  B. albolineatus, 99, **101**
  B. frankei, 99
  B. rerio, 98, 99, **100**
Brichardi, 20, **21**
brine shrimp, 133
Bristlenose Catfish, 105, **123**
Brochis splendens, **114**

Bronze Bubble-eye, **202**
Bronze Cory, **107,** 108
Bubble-eyes, 174, 202–203
Butterfly Barb, 151
Butterflyfish, 159, **160–161**

C

Cactus Pleco, **128–129**
Calico Ryukin, **183**
Callichthyids, 105, 114
Callichthys callichthys, 105
Campylomormyrus tamandua, **169**
Cape Lopez Lyretail, 78, *79*
Carassius
  C. auratus, 174–203
  C. carassius, 176
Cardinal Tetra, 52–53, *53,* **56**
Carnegiella strigata, **67**
catfish, 104–131. *See also Corydoras; Synodontis*
Celebes Rainbowfish, 132, *133,* **136**
Celestial Goldfish, 175
Channa
  C. asiatica, *159,* **164**
  C. micropeltes, **165**
characins, 52–67
Cherry Barb, 151, **155**
Chinese Lionhead, 194
Chocolate Pompon, **198–199**

"Cichlasoma" octofasciatum, **43**
cichlids, 8–51
  African, 8–29, 158
  American, 30–51
Clarias batrachus, 105
Clown Knifefish, **167**
Clown Loach, 138, **143**
Clown Rasbora, 99
Clown Synodontis, **116**
Cobitidae, 138
Cockatoo Dwarf Cichlid, **51**
Colisa
  C. chuna, **96**
  C. fasciata, **95**
  C. lalia, **94, 95,** 96
Combtails, 86
Comet Goldfish, 175, **176–177**
Compressiceps, **10–11**
Compressiceps Dwarf Pike, 36, **37**
Congo Tetra, 53, **63**
Convict Cichlid, 30
Copadichromis, 8
Copper Rosy Barb, 150, 151, **154**
Corydoras, 104–105, **106–113**
  C. adolfoi, 111
  C. aeneus, **106–107,** 108
  C. cf. aeneus, **108**
  C. duplicareus, **111**
  C. gossei, 104, **113**
  C. panda, **110**
  C. sterbai, **112**
Crenicichla
  C. compressiceps, 36, **37**
  C. lepidota, **36**

Crossocheilus
  C. denisonii, **148**
  C. siamensis, 139, 144, **145**
Crown Pearlscale, **201**
Crucian Carp, 176
Cryptoheros nigrofasciatus, 30
Crystal-eye Catfish, 104, **130–131**
Cuckoo Catfish, 105, **117**
Cyphotilapia frontosa, 8–9, 14, **15**
Cyprichromis, 8
  C. sp. 'Leptosoma Jumbo,' **14**
Cyprinidae, 98–99, 150

D

Daffodil, **20**
Danio, 98
  D. aequipinnatus, 99, **100, 101**
  D. devario, 99
danios, 98–101
Delhezi's Bichir, **171**
Dianema urostriata, **115**
Discus, 31, **46–49**
Distochodus, 53
  D. lusosso, **54,** 55
  D. sexfasciatus, **55**
Duboisi, **24**
Duplicate Cory, **111**
Dwarf Cichlids, **50, 51**
Dwarf Gouramis, **94–95**

E

Electric Blue Hap, **22**
Electric Eel, 158, 166
Elephantnose, 158–159, **168**
Emerald Catfish, **114**
Epalzeorhynchos
  E. bicolor, 138, 139, **141**
  E. frenatus, **140, 141**
  E. kalopterus, 139, **144**
Epiplatys, 79
  E. chaperi angona, **83**
  E. roloffi, **83**
Eretmodus cyanosticus, 9, **19**
Erpetoichthys calabaricus, 159, **173**

F

Fantails, 174, **180–181**
Farlowella acus, **122**
Featherfin Syno, 105, **118–119**
Featherfin Tetra, 52
Festive Cichlid, 31
Firemouth Cichlid, 30
Flagtail Catfish, **115**
Flag Tetra, 52
Flower Horn Cichlid, 30
Flying Fox, 139, **144**
Frontosa, 8–9, 14, **15**
Fundulopanchax
  F. gardneri, **80**
  F. gardneri nigerianum, 78, **81**
  F. spoorenbergi, **85**
  F. walkeri, 79

G

Gambusia, 68
Gardner's Aphyosemion, **80**
Gastromyzon sp., 138
Giant Danio, 99, **100, 101**
Giant Gourami, 95, 159
Glossolepis incisus, 132, 133, **137**
Glowlight Tetra, 52, 53, **63**
Gnathonemus petersi, 158–159, **168**
Goby Cichlid, 9, **19**
Gold Danio, 98
Golden Barb, 151
Golden Gourami, **92**
Golden Tetra, 52
goldfish, 174–203
Gold Giant Danio, **101**
Goldy Pleco, **126–127**
Gold Zebra Loach, **146**
Gombi Julie, 8, **18**
Goodeids, 68
Gosse's Cory, 104, **113**
Gouramis, 86, 87, **92–97**
  dwarf, 94, 95, 96
Green Sailfin Molly, 68, 69, **72**
Green Striped Swordtail, **77**

Green Swordtail, 68
Guppies, 68–69, **70–71,** 78

H

Hama Nishiki, **200**
haplochromines, 8
Hard-lipped Barb, 150
Harlequin Rasbora, 99, **102**
hatchetfish, 53, **67**
Head-and-tail-light Tetra, 52
Headstander, 52, 53, **67**
Helostoma temmincki, **92**
Hemibagrus wyckii, 104, **130–131**
Hemichromis spp., 9
  H. guttatus, 16
  H. lifalili, **16**
  H. paynei, **17**
Hemigrammus, 52
  H. erythrozonus, **63**
Heros
  H. efasciatus, 31, **38–39**
  H. severus, 39
Hillstream Loach, 138
Hockey Stick Tetra
  Nannostomus eques, 53
  Thayeria boehlkei, **65**
Honey Gourami, **96**
Hora's Loach, 138, **147**
Hornet Tilapia, 8, 9, **23**
Hypancistrus zebra, **125**
Hyphessobrycon, 52
  H. eques, **64**
  H. erythrostigma, **62**
  H. pulchripinnis, **57**

Hypostomus sp. L124, **124**
Hypsophrys nicaraguensis, 30

I

Ice-blue Zebra, **28**
Indian Glassfish, 109
Iodotropheus sprengerae, **29**
Iriatherina werneri,
132–133, **134**

J

Jack Dempsey, **43**
Jadehead, 196
Jaguar Cichlid, 30, **40–41**
Jewel Cichlids, 9, 16–17
Jordanella floridae, 79, **84**
Julidochromis, 8
  J. sp. 'Gombi,' *8*, **18**

K

killifish, 78–85
King Cobra Guppy, **70**
King Galaxy Pleco, **130**
Kissing Gourami, **92**
knifefish, 159, **166–167**
Kohaku, 196
Krib, 9
Kryptopterus bicirrhis, 105
Kuhli Loach, 138, **147**

L

Labeo chrysophekadion, **149**
Labeotropheus, 8
  L. trewavasae, **18**
Labidochromis, 8
labyrinth fish, 86–97
Ladder Loach, **146**
Lamprologus, 9
  L. ocellatus, **17**
Leleupi, **20**
Lemon Tetra, 52, **57**
Leopard Danio, 99
Lepidota Pike Cichlid, **36**
Lepisosteus oculatus, 159, **172**
Leporinus fasciatus,
  52, 53, **67**
Leptobarbus hoevenii,
  151, **157**

Leptosoma Jumbo, **14**
Lethrinops, 8
live-bearers, 68–77
loaches,
  138–139, 143, 146–147
Long-nosed Distichodus, **54,** 55

M

Madagascar Rainbowfish, 132
Marbled Elephantnose, **169**
Marbled Hatchetfish, **66**
Marosatherina ladigesi,
  132, *133*, **136**
Mbuna, 8, *9*, 12, 27–29
Melanochromis, 8
Melanotaenia
  M. boesmani, **135**
  M. splendida, 134
  M. splendida australis, **134**
Melanotaenidae, 132, 133
Mesonauta insignis, 31
Mesonomacheilus
  triangularis, **147**
Metriaclima, 8, 9
  M. greshakei, **28**
Metynnis, 53
  M. argenteus, **61**
Microgeophagus ramirezi,
  31, **42**
Midas Cichlid, 30, **32–33**
minnows, 98, 102
Mollies, 68–69, 72–73
Moor, 174, **190–193**
Mosquitofish, 68
Myleus, 53
  M. rubripinnis, *52*, **60**
Mylossoma, 53

N

Nannacara, 31
Nannostomus eques, 53
Neolamprologus
  N. brichardi, 20, **21**
  N. leleupi, **20**
  N. leloupi, 20
  N. sp. 'Daffodil,' **20**
Neon Blue Guppy, **70**
Neon Tetra, 52–53, *53*, **56**
Nimbochromis polystigma, 158
Notopterus chitala, **167**

O

Ocellatus, **17**
Opaline Gourami, **97**
Orandas, **184–187**, **200**
Orange Aggie, **51**
Orange Bubble-eye, **203**
Orange Platy, **74**
Orange Sailfin Molly, **73**
Ornate Bichir, **170**
Oscar, 31, **34–35**
Osphronemus gorami, 95, 159
Osteochilus hasselti, 150
Osteoglossiformes, 159
Osteoglossum bicirrhosum,
  158, 159, **162–163**

P

Pachypanchax sakaramyi, **84**
Panaque nigrolineatus,
  104, **131**
Panchax, 79, 83–85
Panda Butterfly, **192**
Panda Catfish, **110**
Pangio, 138
  P. kuhlii sumatranus, **147**
Pantodon buchholzi, 159, **160**

Paracheirodon
  P. axelrodi, 52–53, *53*, **56**
  P. innesi, 52–53, *53*, **56**
Parachromis managuensis,
  30, **40–41**
Paradisefish, 86
Pearl Danio, 99, **101**
Pearl Discus, **47**
Pearl Gourami, **93**
Pearlscale, 20
Pearlscale Oranda, 200
Pelvicachromis, 9
Pelvicachromis pulcher, 9
pencilfish, 53
Penguin Tetra, 53, **65**
Peruvian Green Stripe Cory, **108**
Petricola, **120**
Phenacogrammus interruptus,
  53, **63**
*Phractocephalus
  hemioliopterus*, **128–129**
Pigeon Blood Discus, **47**
Pike Cichlids, 31, **36–37**
pimelodids, 104, 124
Pimelodus pictus, **124**
Pineapple Swordtail, *68*, **76**
Piracuru, 159
Placidochromis, 8
Platies, 68–69, 74–75, 78
  'Varietus,' 69
Poecilia, 68
  P. latipinna, 72
  P. reticulata, 68–69, **70–71**
  P. velifera, *68*, 69, **72–73**
Poeciliidae, 68–73
Polkadot Loach, **146**
polypterids, 170–171, 173

Polypterus
  P. delhezi, **171**
  P. ornatipinnis, **170**
Pompon, **198–199**
Potamotrygon spp., 158
Pseudacanthicus
  P. sp. L25, **128–129**
  P. sp. L282, **130**
Pseudotropheus, 8, 28
  P. zebra, 9
Pterophyllum
  P. altum, 31, **44**
  P. scalare, 31, **45**
Pufferfish, 158
Puntius. See barbs

R

rainbowfish, 132–137
Rainbow Shark, **140**
Ram, 31, **42**
Ranchu, 174, 194–197
Rasbora
  R. heteromorpha, 99, **102**
  R. trilineatus, **103**
rasboras, *98–99,* 99
Red-and-Silver Metallic
  Ranchu, **195–196**
Red-and-Silver Oranda, **185**
Red-and-Silver Ryukin, **182**
Red Belly Piranha, 53, **58–59**
Red Butterfly Moor, **193**
Redcap Oranda, **187**
Red Dwarf Gourami, **94**
Red-finned Cigar Shark,
  151, **157**
Red Honey Gourami, **96**
Red Hook, *52,* **60**
Red Jewel Cichlid, **16**
Red Marlboro Discus, 31, **46**
Red Metallic Oranda, **186**
Red Metallic Ranchu, 195, **197**
Red Snakehead, **165**
Redtail Catfish, **128–129**
Red-tailed Black Shark, *138,*
  139, **141**
Red Torpedo, **148**
Red Wag Swordtail, **77**

Red Wagtail Platy, **75**
Roloff's Panchax, **83**
Ropefish, 159, **173**
Rosy Barb, *150,* 151, **154, 155**
Round-faced Distichodus, **55**
Royal Pleco, *104,* **131**
Ruby Shark, **140**
Rummy-nose Tetra, 52
Rusty Cichlid, **29**
Ryukin, 181, **182–183**

S

Sailfin Mollies, 72–73
Salmon Red Rainbow,
  132, 133, **137**
Schubert's Barb, 151
Sciaenochromis fryeri, **22**
Scissortail Rasbora, *98,* 99, **103**
Scobinancistrus aureatus,
  126–127
Serpae Tetra, 52, **64**
Serrasalmus nattereri, 53, **58–59**
Severum, 31, **38–39**
sharks, 139, 140–142,
  144–145, 148–149, 157
Shubunkin, **178–179**
Siamese Fighting Fish.
  See Betta splendens
Siamese Flying Fox,
  139, 144, **145**
Silver Arowana, *158,* 159, **162–163**
Silver Balloon Molly, 73
Silver Dollar, **61**
Silver Sailfin Molly, **73**
Skunk Loach, 138, **147**
Snakehead, *159,* **164**
Snakeskin Discus, **48**
Splendid Rainbow, **134**
Spoorenberg's Panchax, **85**
Spotted Garpike, *159,* **172**
Spotted Pim, **124**
Steatocranus, 9
Steatocranus casuarius, 9
Steel-blue Lyretail, **81**
Sterba's Cory, **112**
Stingrays, 158
Striped Barb, *150,* **156**

Striped Headstander, 53
Sunrise Guppy, **71**
Sunset Cory, **109**
Swordtails, 68–69, 76–77
Symphysodon
  S. aequifasciatus, 31, **46–49**
  S. discus, 31
Synodontis
  S. angelicus, **117**
  S. decorus, **116**
  S. eupterus, 105, **118–119**
  S. multipunctatus, 105, **117**
  S. nigriventris, 105, **120**
  S. petricola, **120**

T

Talking Catfish, 105
Tanichthys albonubes, 98, **102**
Teleogramma brichardi, 9
Telescope-eye, *175,* **188–189**
Tetraodon biocellatus, 158
tetras, 52–53, 56–57, 62–65,
  132
Thayeria boehlkei, 53, **65**
Thorichthys meeki, 30
Threadfin Rainbowfish,
  132–133, **134**
Three-lined Rasbora, *98,*
  99, **103**
Three-spot Gourami, **97**
Tiger Barb, 139, 150, *151,* **152–153**
Tilapia buttikoferi, 8, 9, **23**
Tinfoil Barb, 150
Trewavas' Cichlid, **18**
Triangle Loach, **147**
Trichogaster, 87
  T. leeri, **93**
  T. trichopterus, **92, 97**
  T. trichopterus trichopterus,
  97

Tropheus
  T. duboisi, **24**
  T. moorii, 26–27
  T. moorii 'Cherry Spot,' **27**
  T. moorii 'Katoto,' **26**
  T. moorii 'Lupota,' **26**
  T. moorii 'Orange,' **27**
Tuxedo Platy, **75,** 76
Twig Catfish, **122**

U

Upside-down Catfish,
  105, **120**
Utaka, 8

V

Veiltails, 187, 190–193
Viejita Dwarf Cichlid, *30,* **50**

W

Walking Catfish, 105
White Cloud Mountain
  Minnow, 98, **102**

X

Xenotilapia, 9
Xiphophorus, 68
  X. helleri, 69, 76–77
  X. maculatus, 69, 74–75

Z

Zebra Danio, 98, 99, **100**
Zebra Pleco, **125**

ACKNOWLEDGMENTS

The publishers would like to thank the following hobbyists and aquatic stores for their invaluable help during the preparation of this book:

David and Paul Cummings at Kesgrave Tropicals, Ipswich, Suffolk.

Dave, Jason and Lewis at Swallow Aquatics, Colchester, Essex.

Richard Hardwick, Ivan Mogford and Wayne at Wharf Aquatics, Pinxton, Nottinghamshire.

Paul Sparks, Chris Dickerson and Mike Laws of the Thorpe Aquarist Society in the Norwich and Kings Lynn area.

Adrian Burge, Publicity Officer of the British Killifish Association.

Glen Bird and Dean at The Aquatic Warehouse, Enfield, Middlesex.

Brian Paddock and David Quelch at Waterworld, Enfield, Middlesex.

Andy Taylor at Britain's Aquatic Superstore, Bolton, Lancashire.

Neil Woodward, Clare, David and John at Pier Aquatics, Wigan, Lancashire.

Gareth Copping and Warren Hall at Shotgate Aquatics, Basildon, Essex.

Peter Hiscock, Steve and Emma at Maidenhead Aquatics, Crowland, Peterborough.

Pete Daniels, Hadleigh, Suffolk.

Colin and Kay Sargeant, Stowmarket, Suffolk.

Rachel, Jamie, Carol and James at Hertfordshire Fisheries, St. Albans, Hertfordshire.

Robin Moore at Knights Ivy Mill, Godstone, Surrey.

Maureen McGurk at Koishi, Tewin, Welwyn, Hertfordshire.

Dick Mills; Dennis Roberts; Derek Lambert.

The photograph of the adult Flower Horn Cichlid shown on page 30 is © Nathan Chiang, Taiwan.

Cichlid consultant: Mary Bailey

North American Consultant: Dominic Stones

The information and recommendations in this book are given without any guarantees on the part of the author and publishers, who disclaim any liability with the use of this material.

And, of course, we must thank the fish, including the one that nearly got away! **Geoff Rogers**